T0136145

Manufacturing and Enterprise
An Integrated Systems Approach

Systems Innovation Series

Series Editor

Adedeji B. Badiru

Air Force Institute of Technology (AFIT) – Dayton, Ohio

Systems Innovation refers to all aspects of developing and deploying new technology, methodology, techniques, and best practices in advancing industrial production and economic development. This entails such topics as product design and development, entrepreneurship, global trade, environmental consciousness, operations and logistics, introduction and management of technology, collaborative system design, and product commercialization. Industrial innovation suggests breaking away from the traditional approaches to industrial production. It encourages the marriage of systems science, management principles, and technology implementation. Particular focus will be the impact of modern technology on industrial development and industrialization approaches, particularly for developing economics. The series will also cover how emerging technologies and entrepreneurship are essential for economic development and society advancement.

Defense Innovation Handbook

Guidelines, Strategies, and Techniques

Edited by Adedeji B. Badiru and Cassie Barlow

Productivity Theory for Industrial Engineering

Ryspek Usubamatov

Quality Management in Construction Projects

Abdul Razzak Rumane

Company Success in Manufacturing Organizations

A Holistic Systems Approach

Ana M. Ferreras and Lesia L. Crumpton-Young

Introduction to Industrial Engineering

Avraham Shtub and Yuval Cohen

Design for Profitability

Guidelines to Cost Effectively Manage the Development Process of Complex Products

Salah Ahmed Mohamed Elmoselhy

Handbook of Measurements

Benchmarks for Systems Accuracy and Precision

Edited by Adedeji B. Badiru, LeeAnn Racz

Global Engineering

Design, Decision Making, and Communication

Carlos Acosta, V. Jorge Leon, Charles R. Conrad and Cesar O. Malave

For more information about this series, please visit: https://www.crcpress.com/Systems-Innovation-Book-Series/book-series/CRCSYSINNOV

Manufacturing and Enterprise
An Integrated Systems Approach

by
Adedeji B. Badiru, Oye Ibidapo-Obe,
and Babatunde J. Ayeni

CRC Press
Taylor & Francis Group
Boca Raton London New York

CRC Press is an imprint of the
Taylor & Francis Group, an **informa** business

CRC Press
Taylor & Francis Group
6000 Broken Sound Parkway NW, Suite 300
Boca Raton, FL 33487-2742

First issued in paperback 2021

© 2019 by Taylor & Francis Group, LLC

CRC Press is an imprint of Taylor & Francis Group, an Informa business
No claim to original U.S. Government works

ISBN-13: 978-0-367-78053-1 (pbk)
ISBN-13: 978-1-4987-4488-1 (hbk)

Library of Congress Cataloging-in-Publication Data

Names: Badiru, Adedeji Bodunde, 1952- author. | Ibidapo-Obe, Oye, author. | Ayeni, Babatunde J., author.
Title: Manufacturing and enterprise: an integrated systems approach / by Adedeji B. Badiru, Oye Ibidapo-Obe, and Babatunde J. Ayeni.
Description: Boca Raton : CRC Press/Taylor & Francis, [2019] | Series: Systems innovation series
Identifiers: LCCN 2018039375| ISBN 9781498744881 (hardback : acid-free paper) | ISBN 9781498744898 (ebook)
Subjects: LCSH: Manufacturing processes.
Classification: LCC TS183 .B33 2019 | DDC 670--dc23
LC record available at https://lccn.loc.gov/2018039375

Visit the Taylor & Francis Web site at
http://www.taylorandfrancis.com

and the CRC Press Web site at
http://www.crcpress.com

Contents

Preface

Manufacturing is the cornerstone of international commerce. Students in engineering programs must learn how manufacturing and the business enterprise work, from an integrated systems approach. This book bridges the traditional gap between manufacturing and business. Manufacturing provides the foundation for a sustainable industrialization. Manufacturing refers to the activities and processes geared toward the production of consumer products. Industrialization refers to the broad collection of activities, services, resources, and infrastructure needed to support business and industry. A nation that cannot institute and sustain manufacturing will be politically delinquent and economically retarded in the long run. A good industrial foundation can positively drive the political and economic processes in any nation. This foundation must start in an academic setting. This textbook contributes to building the desired foundation. Topics covered include engineering processes, systems modeling, business enterprise, forecasting, inventory management, product design, and project management.

Acknowledgment

Dr. Babatunde Ayeni sincerely thanks Professor William Li and Professor Lexin Li for their permissions to include their research work on optimal experimental design in Chapter 12 of this book.

Dr. Oye Ibidapo-Obe hereby acknowledges with gratitude the support of Dr. Peter Ozoh in resourcing for Chapters 6, 7, 8, and 9.

Authors

Adedeji B. Badiru is a professor of Systems Engineering at the Air Force Institute of Technology (AFIT). He was previously professor and head of Systems Engineering and Management at AFIT, professor and department head of Industrial & Information Engineering at the University of Tennessee in Knoxville, and professor of Industrial Engineering and Dean of University College at the University of Oklahoma, Norman. He is a registered professional engineer (PE), a certified project management professional (PMP), a fellow of the Institute of Industrial Engineers, and a fellow of the Nigerian Academy of Engineering. He holds a BS in Industrial Engineering, MS in Mathematics, and MS in Industrial Engineering from Tennessee Technological University, and PhD in Industrial Engineering from the University of Central Florida. He is a prolific author and has won national and international awards for his scholarly, leadership, and professional accomplishments. He is a member of several professional associations and scholastic honor societies.

Dr. O. Ibidapo-Obe, FAS, FAEng, FASCE, OFR, is a Distinguished Professor of Systems Engineering and former Vice Chancellor (2000–2007) at the University of Lagos and the Vice Chancellor of the Federal University Ndufu Alike Ikwo Ebonyi State Nigeria (2011–2016). He was awarded a Bachelor of Science degree in Mathematics in the 1st Class Division by the University of Lagos, Nigeria in 1971; a Master of Mathematics degree in Applied Mathematics with a minor in Computer Science in 1973 and a PhD in Civil Engineering with specialization in Applied Mechanics/Systems in 1976, both from the University of Waterloo, Ontario, Canada. He attended the Round Table at Oxford University in 2003 and also the 2007 MIT Research and Development and Innovations for Development Course, as well as Harvard University Kennedy School of Government in 2013. He is an awardee of "distinguished alumnus" of both the University of Lagos and the University of Waterloo. He is the Pro chancellor of The Technical University (Tech-U) Ibadan and vice chairman of the African University of Science and Technology (AUST) Abuja. He was born on

July 5, 1949, married to Olusola on December 20, 1975, with 4 children and 6 grandchildren.

Professor Ibidapo-Obe's major academic contributions are in the areas of control and information systems including specialist interests in stochastic/optimization problems in engineering, reliability studies; simulation and animation studies (with application to urban transportation, water resources, biomedics, expert systems, etc.). He pioneered studies on stochastic methods in mechanics and the development of computer algorithms applying Martingale concepts to the control of non-linear dynamical systems (1975). He was also part of the pioneering efforts on robotics development and applications, especially in the area of sensors and actuators placement (1981–).

He is a member of the NASRDA (National Space Research and Development Agency) Technology Advisory Committee (1981–) and the African Union AU/NEPAD High-Level African Panel in Engineering Technologies (APET) (2016–). He has published extensively in reputable international journals with some 100 papers after the appointment as Professor.

Babatunde (Babs) J. Ayeni, PhD, is currently the division statistician and advanced quality specialist in the Industrial Adhesives and Tapes Division of 3M Company. Dr. Ayeni received his BS (1980) and MS (1981) in petroleum engineering and his PhD (1984) in statistics, all from the University of Louisiana. He was previously professor and chairman in the Department of Technology at Southern University in New Orleans, Louisiana. He possesses a high degree of expertise in industrial statistics and petroleum engineering with significant application experience in major 3M businesses in such areas as quality, R&D, oil and gas, Six sigma, supply chain, manufacturing, and laboratory. He is a recognized expert in the field of statistics both within and outside 3M. He has excellent customer focus and has been providing leadership on statistical methods since 1988. He has provided project and consulting support to most 3M business units in United States, Europe, Asia, and South America, with major contributions to the polyester film plant in Caserta, Italy, and the development of a feedback controller design that controlled a printing plate process in Sulmona, Italy. He also has a strong research capability as evidenced by his more than 50 technical publications and two books published in 1993 and 2011.

Dr. Ayeni also developed models for oil and gas productions. At 3M Oil & Gas, he developed models for gas condensate reservoirs for oil and gas wells in the US, UK, China, Thailand, Mexico, Indonesia, New Zealand, as well as developing models for many fractured treatments for wells in Colombia, South America. Dr. Ayeni has received several academic honors: Pi Mu Epsilon (Mathematics Honor), Phi Eta Sigma

(Freshman Honor), Tau Beta Pi (Engineering Honor), and Pi Epsilon Tau (Petroleum Engineering Honor). He has served on the editorial advisory boards of the *Institute of Industrial Engineers Journal* as well as *Journal of Operations and Logistics*. Dr. Ayeni is a senior member of the Society of Petroleum Engineers and the American Statistical Association. Dr. Ayeni was a recipient of Federal Government of Nigeria Scholarships leading to BS, MS, and PhD degrees. He has contributed substantially to quality and productivity improvements, cost savings, and product innovations at 3M laboratory, R&D, quality, oil and gas, supply chain, and manufacturing environments. He is a recipient of the 3M Specialty Materials Film Division Quality Achievement Award, 3M Kenneth D. Kotnour Award for Leadership in Statistical Thinking, the Minnesota Institute for Nigerian Development (MIND) Professional Achievement Award, 3M Information Technology Division Pyramid of Excellence Award, and 3M Pollution Prevention Pays (3P) award. Dr. Ayeni also served as quality examiner on the Minnesota Quality Award Board of Examiners. He was profiled in *Tall Drums: Portraits of Nigerians Who are Changing America* (Malaysia: Sungai, 1998). Dr. Ayeni also served as an industrial consultant to the United Nations Transfer of Knowledge Through Expatriate Nationals (TOKTEN) program.

chapter one

Modern and traditional manufacturing processes

Introduction

From a technological standpoint (Raman and Wadke, 2014), manufacturing involves the making of products from raw materials through the use of human labor and resources that include machines, tools, and facilities. It could be more generally regarded as the conversion of an unusable state into a usable state by adding value along the way. For instance, a wood log serves as the raw material for making lumber, which in turn is the raw material to produce chairs. The value added is usually represented in terms of cost and/or time. The term "manufacturing" originates from the Latin word *manufactus*, which means "made by hand." Manufacturing has seen several advances during the last three centuries: mechanization, automation, and, most recently, computerization leading to the emergence of direct digital manufacturing, which is popularly known as 3D printing.

Processes that used to be predominantly done by hand and with hand tools have evolved into sophisticated processes making use of cutting-edge technology and machinery. A steady improvement in quality has resulted with today's specifications even on simple toys exceeding those achievable just a few years ago. Mass production, a concept developed by Henry Ford, has advanced so much that it is now a complex, highly agile, and highly automated manufacturing enterprise. There are several managerial and technical aspects of embracing new technologies to advance manufacturing. Sieger and Badiru (1993), Badiru (1989, 1990, 2005), and others address the emergence and leveraging of past manufacturing-centric techniques of expert systems, flexible manufacturing systems (FMS), nano-manufacturing, and artificial neural networks. In each case, strategic implementation, beyond hype and fad, is essential for securing the much-touted long-term benefits. It is envisioned that the contents of this handbook will expand the knowledge of readers to facilitate a strategic embrace of 3D printing.

What is 3D printing?

3D printing, also known as *additive manufacturing* or *direct digital manufacturing*, is a process for making a physical object from a three-dimensional digital model, typically by laying down many successive thin layers of a material. The successive layering of materials constitutes the technique of additive manufacturing. Thus, the term "direct digital manufacturing" stems from the process of going from a digital blueprint of a product to a finished physical product. Manufacturers can use 3D printing to make prototypes of products before going for full production. In educational settings, faculty and students use this process to make project-related prototypes. Open-source and consumer-level 3D printers allow for creating products at home, thus advancing the concept of distributed additive manufacturing. Defense-oriented and aerospace products are particularly feasible for the application of 3D printing. The military often operate in remote regions of the world, where quick replacement of parts may be difficult to accomplish. With 3D printing, rapid production of routine replacement parts can be achieved onsite at low cost to meet urgent needs. The military civil engineering community is particularly fertile for the application of 3D printing for military asset management purposes.

By all accounts, 3D printing is now energizing the world of manufacturing. The concept of 3D printing was initially developed by Charles W. Hull in the 1980s (US Patent, 1986) as a stereolithography tool for making basic polymer objects. Today, the process is used to make intricate aircraft and automobile components. We are now also seeing more and more applications in making prostheses. The first commercial 3D printing product came out in 1988 and proved a hit among auto manufacturers and aerospace companies. Design of medical equipment has also enjoyed a boost due to 3D printing capabilities. The possibilities appear endless from home-printed implements to the printing of complex parts in outer space. The patent abstract (US Patent, 1986) for 3D printing states the following:

> A system for generating three-dimensional objects by creating a cross-sectional pattern of the object to be formed at a selected surface of a fluid medium capable of altering its physical state in response to appropriate synergistic stimulation by impinging radiation, particle bombardment or chemical reaction, successive adjacent laminae, representing corresponding successive adjacent cross-sections of the object, being automatically formed and integrated together to provide a step-wise laminar buildup of the desired object, whereby a three-dimensional object is formed and drawn from a substantially planar surface of the fluid medium during the forming process.

In order to understand and appreciate the full impact and implications of direct digital manufacturing (3D printing), we need to understand the traditional manufacturing processes that 3D printing is rapidly replacing. Sirinterlikci (2014) presents the various aspects and considerations of manufacturing systems as summarized below.

Definition of manufacturing and its impact on nations

Hitomi (1996) differentiated between the terms "production" and "manufacturing." Production encompasses both making tangible products and providing intangible services while manufacturing is the transformation of raw materials into tangible products. Manufacturing is driven by a series of energy applications, each of which causes well-defined changes in the physical and chemical characteristics of the materials (Dano, 1966).

Manufacturing has a history of several thousand years and may impact humans and their nations in the following ways (Hitomi, 1994):

- *Providing basic means for human existence*: Without the manufacture of products and goods humans are unable to live, and this is becoming more and more critical in our modern society.
- *Creating wealth of nations*: The wealth of a nation is impacted greatly by manufacturing. A country with a diminished manufacturing sector becomes poor and cannot provide a desired standard of living for its people.
- *Moving toward human happiness and stronger world peace*: Prosperous countries can provide better welfare and happiness to their people in addition to stronger security while posing less of a threat to their neighbors and each other.

In 1991, the National Academy of Engineering/Sciences in Washington, D.C. rated manufacturing as one of the three critical areas necessary for America's economic growth and national security, the others being science and technology (Hitomi, 1996). In recent history, nations which became active in lower level manufacturing activities have grown into higher level advanced manufacturing and a stronger research standing in the world (Gallagher, 2012).

As the raw materials are converted into tangible products by manufacturing activities, the original value (monetary worth) of the raw materials is increased (Kalpakjian and Schmid, 2006). Thus, a wire coat hanger has a greater value than its raw material, the wire. Manufacturing activities may produce *discrete products* such as engine components, fasteners, and gears, or *continuous products* like sheet metal, plastic tubing, and conductors that are later used in making discrete products. Manufacturing occurs in a complex environment that connects multiple activities: product design, process planning and tool engineering, materials engineering, purchasing and receiving, production control, marketing and sales, shipping, and customer support services (Kalpakjian and Schmid, 2006).

Manufacturing processes and process planning

Manufacturing processes

Today's manufacturing processes are extensive and continuously expanding while presenting multiple choices for manufacturing a single part of a given material (Kalpakjian and Schmid, 2006). The processes can be classified as traditional and non-traditional before they can be divided into their mostly physics-based categories. While much of the traditional processes have been around for a long time, some of the non-traditional processes may have been in existence for some time as well, such as in the case of electro-discharge machining, but not utilized as a controlled manufacturing method until a few decades ago.

Traditional manufacturing processes can be categorized as:

1. Casting and molding processes
2. Bulk and sheet forming processes
3. Polymer processing
4. Machining processes
5. Joining processes
6. Finishing processes

Non-traditional processes include:

1. Electrically based machining
2. Laser machining
3. Ultrasonic welding
4. Water-jet cutting
5. Powder metallurgy
6. Small scale manufacturing
7. Additive manufacturing
8. Biomanufacturing

Process planning and design

The selection of a manufacturing process or a sequence of processes depends on a variety of factors including the desired shape of a part and its material properties for performance expectations (Kalpakjian and Schmid, 2006). Mechanical properties such as strength, toughness, ductility, hardness, elasticity, fatigue, and creep; physical properties such as density, specific heat, thermal expansion and conductivity, melting point, magnetic and electrical properties as well as chemical properties such as oxidation, corrosion, general degradation, toxicity, and flammability may play a major role in the duration of the service life of a part and recyclability. The manufacturing properties of materials are also critical since they determine whether the material can be cast, deformed, machined, or heat treated into the desired shape. For example, brittle and hard materials cannot be deformed without failure or high energy requirements; whereas they cannot be machined unless a non-traditional method such electro-discharge machining is employed. Table 1.1 depicts general manufacturing characteristics of various alloys and can be utilized in the selection of processes based on the material requirements of parts.

Each manufacturing process has its characteristics, advantages, and constraints including production rates and costs. For example, the conventional blanking and piercing process used in making sheet metal parts can be replaced by its laser-based counterparts if the production rates and costs can justify such a switch. Eliminating the need for tooling will also be a plus as long as the surfaces delivered by the laser-cutting process is comparable or better than that of the conventional method (Kalpakjian and Schmid, 2006). Quality is a subjective metric in general (Raman and Wadke, 2014); however, in manufacturing it often implies *surface finish and tolerances, both dimensional and geometric*. The economics of any process is again very important and can be conveniently decomposed with the analysis of manufacturing operations and their tasks. A manufactured part can be broken into its features, the features can be meshed with certain operations, and operations can be separated into their tasks. Since several possible operations may be available and multiple sequences of operations co-exist, several viable process plans can be made (Raman and Wadke, 2014).

Process routes are a sequence of operations through which raw materials are converted into parts and products. They must be determined after

Table 1.1 Amenability of alloys for manufacturing processes

Type of alloy	Amenability for		
	Casting	Welding	Machining
Aluminum	Very High	Medium	High to Very High
Copper	Medium to High	Medium	Medium to High
Gray cast iron	Very High	Low	High
White cast iron	High	Very Low	Very Low
Nickel	Medium	Medium	Medium
Steels	Medium	Very High	Medium
Zinc	Very High	Low	Very High

the completion of production planning and product design according to the conventional wisdom (Hitomi, 1996). However, newer concepts like concurrent or simultaneous engineering or Design for Manufacture and Assembly (DFMA) are encouraging simultaneous execution of part and process design and planning processes and additional manufacturing-related activities. Process planning includes the two basic steps (Timms and Pohlen, 1970):

1. *Process design* is macroscopic decision-making for an overall process route for the manufacturing activity.
2. *Operation/Task design* is microscopic decision-making for individual operations and their detail tasks within the process route.

The main problems in process and operation design are: analysis of the work-flow (flow-line analysis) for the manufacturing activity and selecting the workstations for each operation within the work-flow (Hitomi, 1996). These two problems are interrelated and must be resolved at the same time. If the problem to be solved is for an existing plant, the decision is made within the capabilities of that plant. On the contrary, an optimum work-flow is determined, and then the individual workstations are developed for a new plant within the financial and physical constraints of the manufacturing enterprise (Hitomi, 1996).

Work-flow is a sequence of operations for manufacturing activity. It is determined by manufacturing technologies and forms the basis for operation design and layout planning. Before an analysis of work-flow is completed, certain factors have to be defined including precedence relationships and work-flow patterns. There are two possible relationships between any two operations of the work-flow (Hitomi, 1996):

1. A partial order, *precedence*, exists between two operations such as in the case of counterboring. Counterboring must be conducted after drilling.
2. No precedence exists between two operations if they can be performed in parallel or concurrently. Two sets of holes with different sizes in a part can be made in any sequence or concurrently.

Harrington (1973) identifies three different work-flow patterns: *sequential* (tandem) process pattern of gear manufacturing, *disjunctive* (decomposing) pattern of coal or oil refinery processes, and *combinative* (synthesizing) process pattern in assembly processes.

According to Hitomi (1996), there are several alternatives for work-flow analysis depending on production quantity (demand volume, economic lot size), existing production capacity (available technologies, degree of automation), product quality (surface finish, dimensional accuracy and tolerances), raw materials (material properties, manufacturability). The best work-flow is selected by evaluating each alternative based on a criterion that minimizes *the total production (throughput) time,* or *total production cost,* defined in 1. and 2. above. *Operation process or flow process charts* can be used to define and present information for the work-flow of the manufacturing activity. Once an optimum work-flow is determined, the detail design process of each operation and its tasks are conducted. A *break-even analysis* may be needed to select the right equipment for the workstation. Additional tools such as *man-machine analysis* as well as *human factors analysis* are also used to define the details of each operation. *Operation sheets* are another type of tool used to communicate about the requirements of each task making up individual operations.

$$\text{Total production time} = \sum \big[\text{Transfer time between stages} + \text{waiting time}$$

$$+ \text{set-up time} + \text{operation time} + \text{inspection time} \big]$$

$$\text{Total production cost} = \text{Material cost} + \sum [\text{cost of transfer between stages}$$

$$+ \text{set-up cost} + \text{operation cost} + \text{tooling cost}$$

$$+ \text{inspection cost} + \text{work-in-process inventory cost}]$$

where:

$\sum$ represent all stages of the manufacturing activity.

Industrial engineering and operations management tools have been used to determine optimum paths for the work-flow. Considering the amount of effort involved in the complex structure of today's manufacturing activities, computer-aided process planning (CAPP) systems have become very attractive in generating feasible sequences and minimizing the lead-time and non-value-added costs (Raman and Wadke, 2014).

Traditional manufacturing processes

Casting and molding processes

Casting and molding processes can be classified into four categories (Kalpakjian and Schmid, 2006):

1. Permanent mold based: Permanent molding, (high and low pressure) die-casting, centrifugal casting, and squeeze casting.
2. Expandable mold and permanent pattern based: Sand casting, shell mold casting, and ceramic mold casting.

3. Expandable mold and expandable pattern: Investment casting, lost foam casting, and single-crystal investment casting.
4. Other processes: Melt-spinning.

These processes can be further classified based on their molds: permanent and expandable mold-type processes (Raman and Wadke, 2014). The basic concept behind these processes is to superheat a metal or metal alloy beyond its melting point or range, then pour or inject it into a die or mold and allow it to solidify and cool within the tooling. Upon solidification and subsequent cooling, the part is removed from the tooling and finished accordingly. Figures 1.1 through 1.9 show illustrations of the various manufacturing processes described in the sections that follow.

The expandable mold processes destroy the mold during the removal of the part or parts such as in sand casting and investment casting. Investment casting results in better surface finishes and tighter tolerances than sand casting. Die-casting and centrifugal casting processes also result

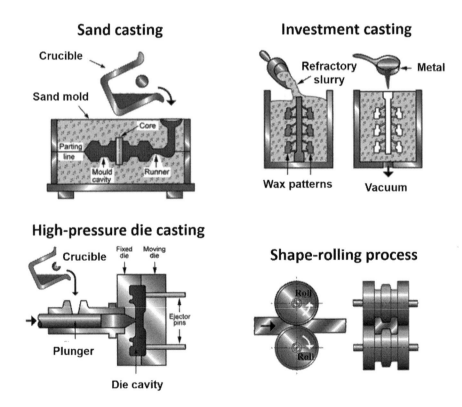

Figure 1.1 Manufacturing processes illustration 1.

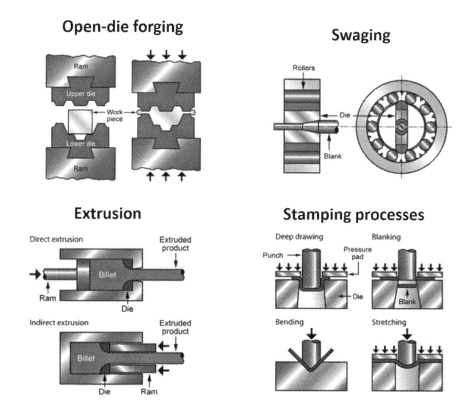

Figure 1.2 Manufacturing processes illustration 2.

in good finishes but are permanent mold processes. In these processes, the preservation of tooling is a major concern since they are reused over and over, sometimes for hundreds of thousands of parts (Raman and Wadke, 2014). Thermal management of tooling through spraying and cooling channels is also imperative since thermal fatigue is a major failure mode for this type of tooling (Sirinterlikci, 2000).

Common materials that are cast include metals such as aluminum, magnesium, copper, and their low-melting-point alloys including zinc, cast iron, and steel (Raman and Wadke, 2014). The tooling used have simple ways of introducing liquid metal, feeding it into the cavity, with mechanisms to exhaust air or gas entrapped within, as well, to prevent defects such as shrinkage porosity and promote solid castings and easy removal of the part or parts. Cores and cooling channels are also included for making voids in the parts and controlled cooling of them to reduce cycle times for the process respectively. The die or mold design, metal fluidity, and solidification patterns are all critical to obtain high-quality

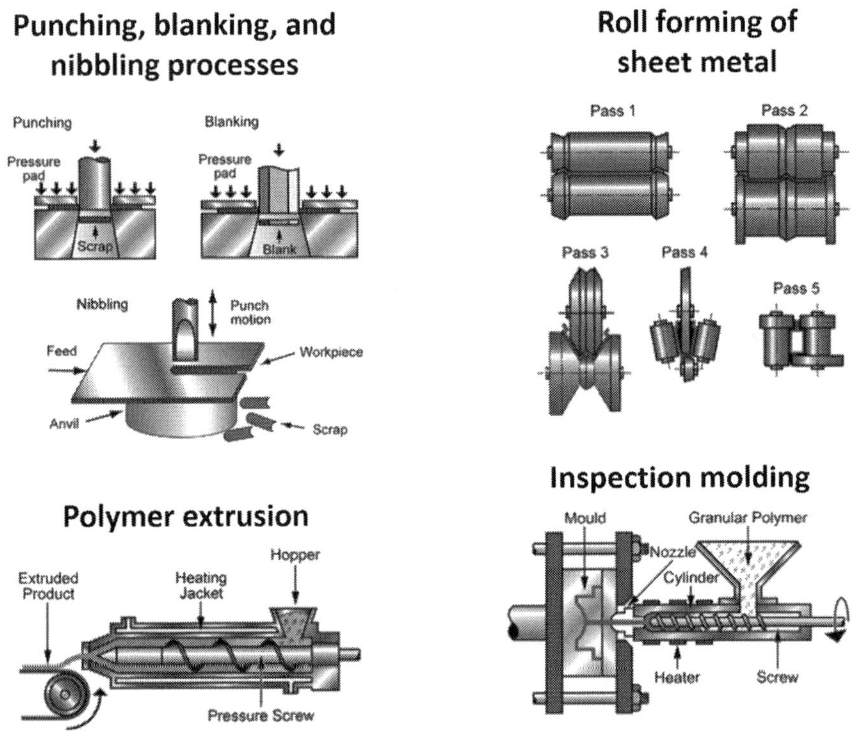

Punching, blanking, and nibbling processes

Roll forming of sheet metal

Polymer extrusion

Inspection molding

Figure 1.3 Manufacturing processes illustration 3.

castings. Suitable provisions are made through allowances to compensate for shrinkage and finishing (Raman and Wadke, 2014).

Bulk forming processes

Forming processes include bulk-metal forming as well as sheet metal operations. No matter what the type or nature of the process, forming applies mainly to metals that are workable by plastic deformation. This constraint makes brittle materials not eligible for forming. Bulk forming is the combined application of temperature and pressure to modify the shape of a solid object (Raman and Wadke, 2014). While cold-forming processes conducted near room temperature require higher pressures, hot working processes take advantage of the decrease in material strength. Consequently, pressure and energy requirements are also much lower in hot processes, especially when the material is heated above its recrystallization temperature, which is 60% of the melting point (Raman and Wadke, 2014). Net shape and near net shape processes accomplish part

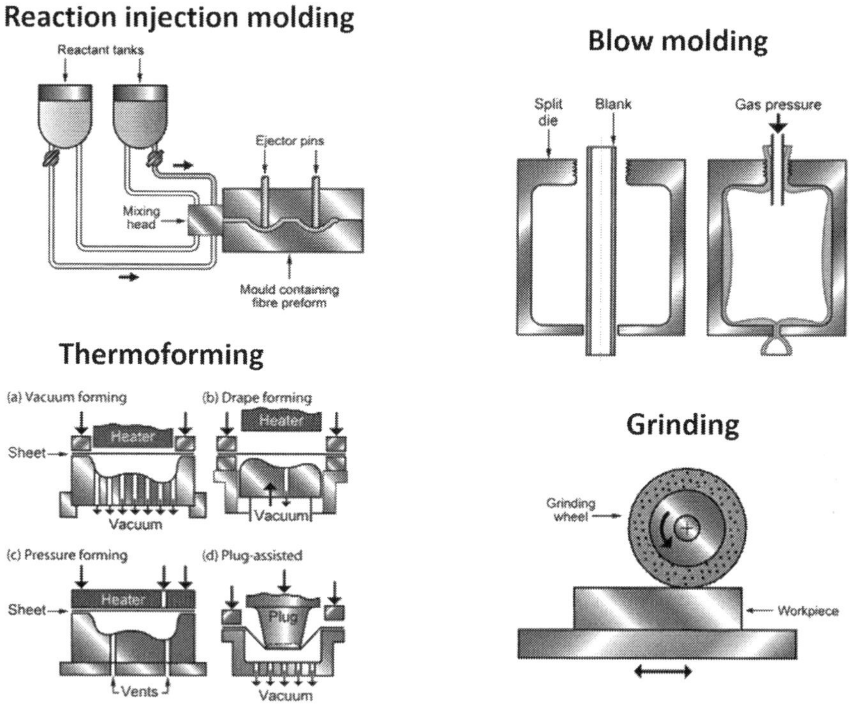

Reaction injection molding

Blow molding

Thermoforming

Grinding

Figure 1.4 Manufacturing processes illustration 4.

dimensions that are exact or close to specification requiring little or no secondary finishing operations.

A group of operations can be included in the classification of bulk forming processes (Kalpakjian and Schmid, 2006):

1. Rolling processes: Flat rolling, shape rolling, ring rolling, and roll forging
2. Forging processes: Open-die forging, closed-die forging, heading, and piercing
3. Extrusion and drawing processes: Direct extrusion, cold extrusion, drawing, and tube drawing

In the flat rolling process, two rolls rotating in opposite directions are utilized in reducing the thickness of a plate or sheet metal. This thickness reduction is compensated by an increase in the length; when the thickness and width are close in dimension, an increase in both width and length occurs based on the preservation of the volume of the parts (Raman and Wadke, 2014). A similar process, shape rolling, is used for obtaining different shapes or cross sections. Forging is used for shaping objects in a

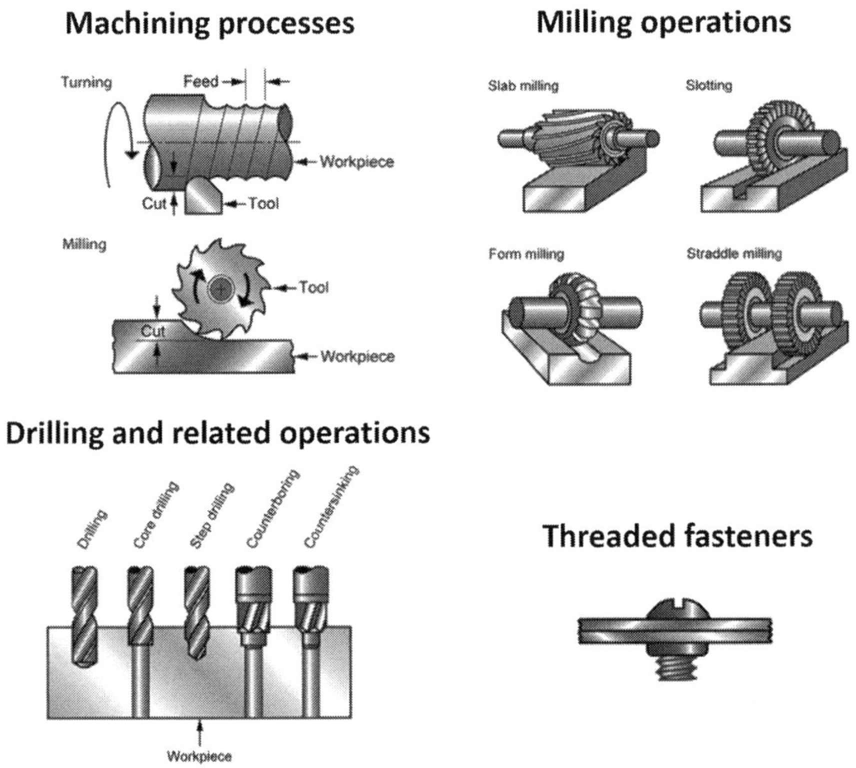

Machining processes

Turning

Feed

Workpiece

Cut

Tool

Milling

Tool

Cut

Workpiece

Milling operations

Slab milling

Slotting

Form milling

Straddle milling

Drilling and related operations

Drilling

Core drilling

Step drilling

Counterboring

Countersinking

Workpiece

Threaded fasteners

Figure 1.5 Manufacturing processes illustration 5.

press and additional tooling, and may involve more than one pre-forming operation, including blocking, edging, and fullering (Raman and Wadke, 2014). Open-die forging is done on a flat anvil and the closed-die forging process uses a die with a distinct cavity for shaping. Open-die forging is less accurate but can be used in making extremely large parts due to its ease on pressure and consequent power requirements. While mechanical hammers deliver sudden loads, hydraulic presses apply gradually increasing loads. Swaging is a rotary variation of the forging process, where a diameter of a wire is reduced by the reciprocating movement of one or two opposing dies. Extrusion is the forcing of a billet out of a die opening similar to squeezing toothpaste out of its container, either directly or indirectly. This process enables fabrication of different cross sections on long pieces (Raman and Wadke, 2014). In co-extrusion, two different materials are extruded at the same time and bond with each other. On the contrary, drawing process is based on pulling of a material through an orifice to reduce the diameter of the material.

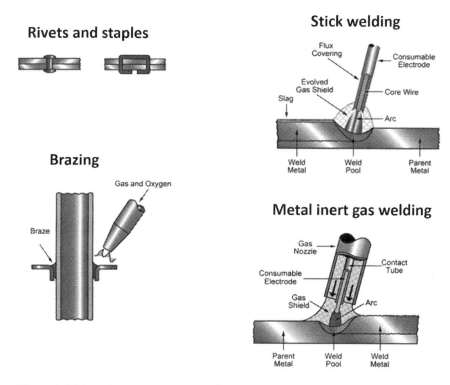

Figure 1.6 Manufacturing processes illustration 6.

Sheet forming processes

Stamping is a generic term used for sheet metal processes. This includes processes such as blanking, punching (piercing), bending, stretching, deep drawing, bending, and coining. Processes are executed singularly or consecutively to obtain complex sheet metal parts with a uniform sheet metal thickness. Progressive dies allow multiple operations to be performed at the same station. Since these tooling elements are dedicated, their costs are high and expected to perform without failure for the span of the production.

Sheet metal pieces are profiled by a number of processes based on shear fracture of the sheet metal. In punching, a shaped hole is obtained by pushing the hardened die (punch) through the sheet metal. In a similar process called perforating, a group of punches are employed in making a hole pattern. In the blanking process, the aim is to keep the part that is punched out, not the rest of the sheet metal with the punched hole. In the nibbling process, a sheet supported by an anvil is cut to a shape by successive bites of a punch similar to the motion of a sewing machine head.

Tungsten inert gas welding

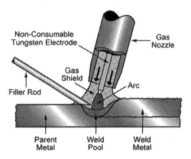

Oxy-fuel welding

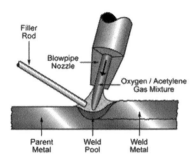

Electroplating

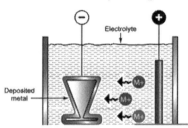

Physical vapor deposition

Figure 1.7 Manufacturing processes illustration 7.

After the perforation of a sheet metal part, different stamping operations may be applied to it. In simple bending, the punch bends the blank on a die. Stretching may be accomplished while tightly holding the sheet metal piece by the pressure pads and forming it in a die. In deep drawing the sheet metal piece is allowed to be deeply drawn into the die while being held by the pressure pads. More sophisticated geometries can be obtained by roll forming in a progressive setting.

Polymer processes

Polymer extrusion is a process to make semi-finished polymer products such as rods, tubes, sheets, and film in mass quantities. Raw materials such as pellets or beads are fed into the barrel to be heated and extruded at temperatures as high as 370°C (698°F). The extrusion is then air or water cooled and may later be drawn into smaller cross sections. Variations of this process are film blowing, extrusion blow molding, and filament forming. This extrusion process is used in the making of blended polymer pellets and becomes a post process for other processes

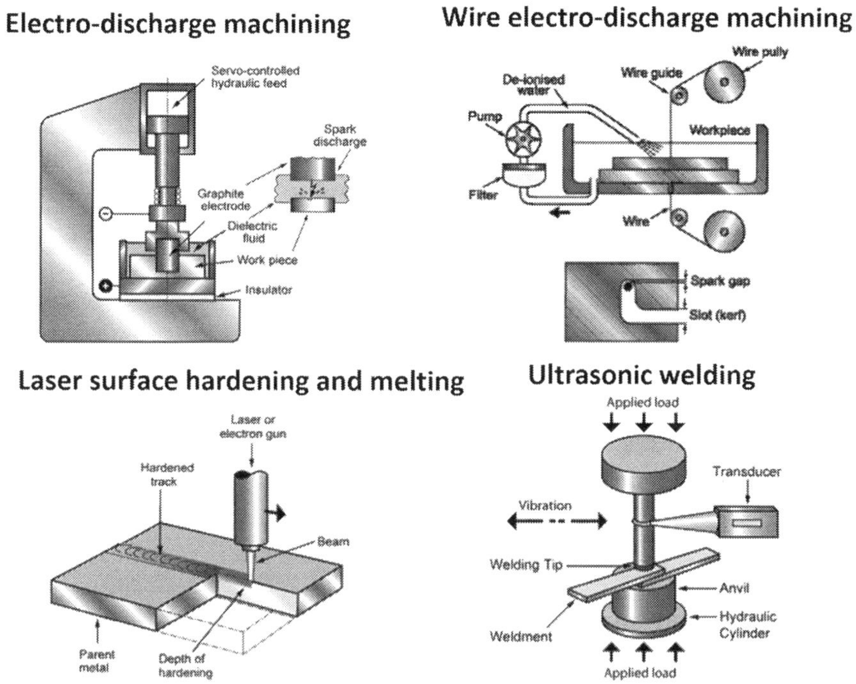

Electro-discharge machining

Wire electro-discharge machining

Laser surface hardening and melting

Ultrasonic welding

Figure 1.8 Manufacturing processes illustration 8.

such as injection molding. It is also utilized in coating metal wires at high speeds.

Injection molding is a process similar to polymer extrusion with one main difference: the extrusion is being forced into a metal mold for solidification under pressure and cooling. The feeding system into the mold includes the sprue area, runners, and gate. Thermoplastics, thermosets, elastomers, and even metal materials are being injection molded. The co-injection process allows molding of parts with different material properties including colors and features. While injection foam molding with inert gas or chemical blowing agents results in large parts with solid skin and cellular internal structure, reaction injection molding (RIM) mixes low-viscosity chemicals under low pressures (0.30–0.70 MPa) (43.51–101.53 psi) to be polymerized via a chemical reaction inside a mold. The RIM process can produce complex geometries and works with thermosets such as polyurethane or other polymers such as nylons and epoxy resins. The RIM process is also adapted to fabricate fiber-reinforced composites.

Adapted from glass blowing technology, the blow molding process utilizes hot air to push the polymer against the mold walls to be frozen.

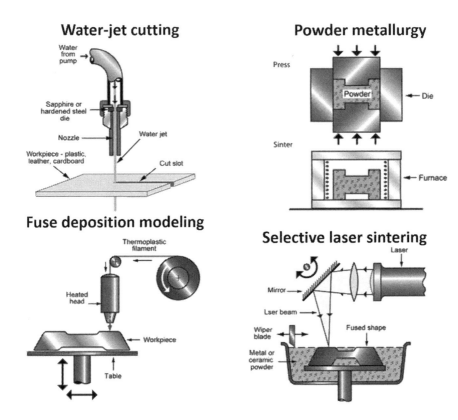

Figure 1.9 Manufacturing processes illustration 9.

The process has multiple variations including extrusion and stretch blow molding. The generic blow molding process allows inclusion of solid handles and has better control over the wall thickness compared with its extrusion variant.

Thermoforming processes are used in making large sheet-based moldings. Vacuum thermoforming applies vacuum to draw the heated and softened sheet into the mold surface to form the part. Drape thermoforming takes advantage of the natural sagging of the heated sheet in addition the vacuum, whereas plug-assisted variant of thermoforming supplements the vacuum with a plug by pressing on the sheet. In addition, pressure thermoforming applies a few atmospheres (atm) of pressure to push the heated sheet into the mold. Various molding materials employed in the thermoforming processes included wood, metal, and polymer foam. A wide variety of other polymer processing methods available include, but not limited to, rotational molding, compression molding, and resin transfer molding (RTM).

Machining processes

Machining processes use a cutting tool to remove chips from the work-piece (Raman and Wadke, 2014). The cutting process requires plastic deformation and the consequent fracture of the work-piece material. The type of chip impacts both the removal of the material and the quality of surface generated. The size and type of the chip are dependent on the type of machining operation and cutting parameters. The chip types are continuous, discontinuous, continuous with a built-up edge, and serrated (Raman and Wadke, 2014). The critical cutting parameters include the cutting speed (in revolutions per minute [rpm], surface feet per minute [sfpm] or millimeters per minute [mm/min]), the feed rate (inches per minute [ipm] or millimeters per minute [mm/min]), and the depth of cut (inches [in] or millimeters [mm]). These parameters affect the work-piece, the tool, and the process itself (Raman and Wadke, 2014). The conditions of the forces, stresses, and temperatures of the cutting tool are deter-mined by these parameters. Typically, the work-piece or tool are rotated or translated such that there is relative motion between the two. A primary zone of deformation causes shearing a material, separating a chip from the work-piece. A secondary zone is also developed based on the friction between the chip and cutting tool (Raman and Wadke, 2014). While rough machining is an initial process to obtain the desired geometry without accurate dimensions and surface finish, finish machining is a precision process capable of great dimensional accuracy and surface finish. Besides metals, stones, bricks, wood, and plastics can be machined.

There are usually three types of chip-removal operations: single-point, multipoint–fixed geometry, and multipoint–random geometry (Raman and Wadke, 2014). Random geometry multipoint operation is also referred to as abrasive machining process, and includes operations such as grinding, honing, and lapping. The cutting tool in a single-point operation resembles a wedge, with several angles and radii to aid cutting. The cutting-tool geometry is characterized by the rake angle, lead or main cutting-edge angle, nose radius, and edge radius. Common single-point operations include turning, boring, and facing (Raman and Wadke, 2014). Turning is performed to make round parts, facing makes flat features, and boring fabricates non-standard diameters and internal cylindrical surfaces. Fixed geometry multipoint operations include milling and drill-ing. Milling operations can be categorized into face milling, peripheral (slab) milling, and end milling. The face milling uses the face of the tool, while slab milling uses the periphery of the cutter to generate the cutting action. These are typically applied to make flat features at a rate of mate-rial removal significantly higher than single-point operations like shap-ing and planning (Raman and Wadke, 2014). End milling cuts along with both the face and periphery and is used for making slots and extensive contours. Drilling is used to make standard-sized holes with a cutter with multiple active cutting edges (flutes). The rotary end of the cutter is used in the material removal process. Drilling has been the fastest and most economical method of making holes into a solid object. A multitude of drilling and relevant operations are available including core drilling, step (peck) drilling, counterboring, and countersinking as well as reaming and tapping.

Assembly and joining processes

Joining processes are employed in the manufacture of multi-piece parts and assemblies (Raman and Wadke, 2014). These processes encompass mechanical fastening through removable bolting and non-removal, adhesive bonding, and welding processes.

Welding processes use different heat sources to cause localized melting of the metal parts to be joined or the melting of a filler to develop a joint between two metals – also heated. Welding of plastics has also been established. Cleaned surfaces are joined together through a butt weld or a lap weld, although other configurations are also feasible (Raman and Wadke, 2014). Two other joining processes are brazing and soldering, which differ from each other in the process temperatures and are not as strong as welding.

Arc welding utilizes an electric arc between two electrodes to generate the required heat for the process. One electrode is the plate to be joined while the other electrode is the consumable one. Stick welding is the most common; it is also called shielded metal arc welding (SMAW). Metal inert gas (MIG) or gas metal arc welding uses a consumable electrode (Raman and Wadke, 2014) as well. The electrode provides the filler and the inert gas provides an atmosphere such that contamination of the weld pool is prevented and, consequently, weld quality is obtained. A steady electrode flow is accomplished automatically to maintain the arc gap, sequentially controlling the temperature of the arc (Raman and Wadke, 2014). Gas tungsten arc welding, or tungsten inert gas (TIG) welding, uses a non-consumable electrode and filler is required for the welding. In resistance welding, the resistance is generated by the air gap between the surfaces to obtain and maintain the flow of electric current between two fixed electrodes. The electrical current is then used to generate the heat required for welding (Raman and Wadke, 2014). This process results in spot or seam welds. Gas welding typically employs acetylene (fuel) and oxygen (catalyzer) to develop different temperatures to heat work-pieces or fillers for welding, brazing, and soldering. If the acetylene is in excess, a reducing (carbonizing) flame is obtained. The reducing flame is used in hard-facing or backhand pipe welding operations. On the contrary, if oxygen is in excess, then an oxidizing flame is generated. The oxidizing flame is used in braze-welding and welding of brasses or bronzes. Finally, if equal proportions of the two are used, a neutral flame results. The neutral flame is used in welding or cutting. Other solid-state processes include thermite welding, ultrasonic welding, and friction welding.

Finishing processes

Finishing processes include surface treatment processes and material removal processes such as polishing, shot-peening, sand blasting, cladding and electroplating, and coating and painting (Raman and Wadke, 2014). Polishing involves very little material removal and is also classified under machining operations. Shot and sand blasting are used to improve surface properties and cleanliness of parts. Chemical vapor deposition (CVD) or physical vapor deposition (PVD) methods are also applied to improve surface properties (Kalpakjian and Schmid, 2006). Hard coatings such as CVD or PVD are applied to softer substrates to improve wear resistance while retaining fracture resistance (Raman and Wadke, 2014) as in the case of metal-coated polymer injection molding inserts (Noorani, 2006). The coatings are less than 10-mm thick in many cases. On the other hand, cladding is done as in aluminum cladding on stainless steel to improve its heat conductivity for thermally critical applications (Raman and Wadke, 2014).

Non-traditional manufacturing processes

There are many non-traditional manufacturing processes. These processes include:

1. Electrically Based Machining: These processes include the electro-discharge machining (EDM), and electrochemical machining (ECM). In the plunge EDM process, the work-piece is held in a work-holder submerged in a dielectric fluid. Rapid electric pulses are discharged between the graphite electrode and the work-piece, causing plasma to erode the work-piece. The dielectric fluid then carries the debris. The wire EDM uses mainly brass wire in place of the graphite electrode, but functions in a similar way. The ECM (also called reverse electroplating) process is similar to the EDM process, but it does not cause any tool wear nor can any sparks be seen. Both processes can be used in machining very hard materials that are electrically conductive.
2. Laser Machining: Lasers are used in a variety of applications ranging from cutting of complex 3D contours (i.e., today's coronary stents) to etching or engraving patterns on rolls for making texture on rolled parts. Lasers are also effectively used in hole-making,

precision micro-machining, removal of coating and ablation. Laser transformation and shock-hardening processes make work-piece surfaces very hard while the laser surface melting process produces refined and homogenized microstructures.

3. Ultrasonic Welding: Ultrasonic welding processes require an ultrasonic generator, a converter, a booster, and a welding tool. The generator converts 50 Hz into 20 KHz. These higher frequency signals are then transformed into mechanical oscillations through reverse piezoelectric effect. The booster and the welding tool transmit these oscillations into the welding area causing vibrations of 10–30 μm in amplitude. Meanwhile, a static pressure of 2–15 MPa is applied to the work-pieces as they slide, get heated, and bonded.

4. Water-Jet Cutting: Water-jet cutting is a machining process employing abrasive slurry in a jet of water to machine hard-to-machine materials (Raman and Wadke, 2014). Water is pumped at high pressures (like 400 MPa) and can reach speeds of 850 m/s (3,060 km/hr or 1901.4 mph). Abrasive slurries are not needed when cutting softer materials.

5. Powder Metallurgy: Powder-based fabrication methods are critical in employing materials with higher melting points due to the hardship of casting them. Once compressed under pressures using different methods and temperatures, compacted (green) powder parts are sintered (fused) usually at 2/3 of their melting points. Common powder metallurgy materials are ceramics and refractory metals, stainless steel, and aluminum.

6. Small Scale Manufacturing: The past two decades have seen the marriage of micro-scale electronics devices with mechanical systems – leading to the design and manufacture of micro-electromechanical devices (MEMS). Newer cutting-edge technologies have emerged in even a smaller scale such as nanotechnology or molecular manufacturing (Raman and Wadke, 2014). The ability to modify and construct products at a molecular level makes nanotechnology very attractive and usable. Single-wall nanotubes are one of the biggest innovations for building future transistors and sensors. Variants of the nano area include nano-manufacturing, such as ultra-high precision machining or adding ionic antimicrobials to biomedical devices (Raman and Wadke, 2014; Sirinterlikci et al., 2012).

7. Additive Manufacturing: Additive manufacturing has developed from additive rapid prototyping technology. Since the late 1980s, rapid prototyping technologies have intrigued scientists and engineers. In the early years, there were attempts of obtaining 3D geometries using various layered approaches as well as direct 3D geometry generation by robotic plasma spraying. Today, a few processes such as fused deposition modeling (FDM), stereolithography

(SLA), laser sintering processes (selective or direct metal), and 3D printing have become household names. There are also other very promising processes such as Objet's Polyjet (Inkjet) 3D printing technology. In the last two decades, the rapid prototyping technology has seen an increase in the number of materials available for processing; layer thicknesses have become thinner while control systems have improved the accuracy of the parts. The end result is shortened cycle times and better quality functional parts. In addition, there have been many successful applications of rapid tooling and manufacturing.

8. Biomanufacturing: Biomanufacturing may encompass biological and biomedical applications. Thus, manufacturing of human vaccinations may use biomass from plants while biofuels are extracted from corn or other crops. Hydraulic oils, printer ink technology, paints, cleaners, and many other products are taking advantage of the developments in biomanufacturing (Sirinterlikci et al., 2010). On the other hand, biomanufacturing is working with nanotechnology, additive manufacturing, and other emerging technologies to improve the biomedical engineering field (Sirinterlikci et al., 2012).

Manufacturing systems

The manufacturing systems are employed to manufacture products and the parts assembled into those parts (Groover, 2001). A manufacturing system is the collection of equipment, people, information, processes, and procedures to realize the manufacturing targets of an enterprise. Groover (2001) defines manufacturing systems in two parts:

1. Physical facilities include the factory, the equipment within the factory, and the way the equipment is arranged (the layout).
2. Manufacturing support systems are a set of company procedures to manage its manufacturing activities and to solve technical and logistics problems including product design and business functions.

Production quantity is a critical classifier of a manufacturing system while the way the system operates is another classifier that includes *production to order* (i.e., Just in Time – JIT manufacturing) and *production for stock* (Hitomi, 1996). According to (Groover, 2001) and (Hitomi, 1996), a manufacturing system can be categorized into three groups based on its production volume:

1. Low production volume – Jobbing systems: In the range of 1–100 units per year, associated with a job shop, fixed position layout, or process layout.

2. Medium production volume – Intermittent or batch systems: With the production range of 100–10,000 units annually, associated with a process layout or cell.
3. High production volume – Mass manufacturing systems: 10,000 units to millions per year, associated with a process or product layout (flow-line).

Each classification is associated inversely with a product variety level. Thus, when product variety is high, production quantity is low, and when product variety is low, production quantity is high (Groover, 2001). Hitomi (1996) states that only 15% of the manufacturing activities in the late 1990s were coming from mass manufacturing systems, while small batch-multi product systems had a share more than 80%, perhaps due to diversification of human demands.

Interchangeability and assembly operations

While each manufacturing process is important for fabrication of a single part, assembling these piece parts into assemblies and subassemblies also present a major challenge. The concept that makes assembly a reality is interchangeability, which relies on standardization of products and processes (Raman and Wadke, 2014). Besides facilitating easy assembly, interchangeability also enables easy and affordable replacement of parts within subassemblies and assemblies (Raman and Wadke, 2014).

The key factors for interchangeability are "tolerances" and "allowances" (Raman and Wadke, 2014). Tolerance is the permissible variation of geometric features and part dimensions on manufactured parts with the understanding that perfect parts are hard to be made, especially repeatedly. Even if the parts could be manufactured perfectly, current measurement tools and systems may not be able to verify their dimensions and features accurately (Raman and Wadke, 2014). Allowances determine the degree of looseness or tightness of a fit in an assembly of two parts (i.e., a shaft and its bearing) (Raman and Wadke, 2014). Depending on the allowances, the fits are classified into "clearance," "interference," and "transition" fits. Since most commercial products and systems are based on assemblies, tolerances (dimensional or geometric) and allowances must be suitably specified to promote interchangeable manufacture (Raman and Wadke, 2014).

Systems metrics and manufacturing competitiveness

According to Hitomi (1996), efficient and economical execution of manufacturing activities can be achieved by completely integrating the material flow (manufacturing processes and assembly), information flow (manufacturing management system), and cost flow (manufacturing economics). A manufacturing enterprise needs to serve for the welfare of the society by not harming the people and the environment (green manufacturing, environmentally conscious manufacturing) along with targeting its profit objectives (Hitomi, 1996). A manufacturing enterprise needs to remain competitive and thus has to evaluate its products' values and/or the effectiveness of their manufacturing system by using the following three metrics (Hitomi, 1996):

1. Function and quality of products
2. Production costs and product prices
3. Production quantities (productivity) and on-time deliveries
4. Following industry regulations

A variety of means are needed to support these metrics including process planning and control, quality control, costing, safety, health and environment (SHE), production planning, scheduling, and control including assurance of desired cycle/takt and throughput times. Many other metrics are also used in the detail design and execution of systems including machine utilization, floor space utilization, and inventory turnover rates.

Automation of systems

Some parts of a manufacturing system need to be automated while other parts remain under manual or clerical control (Groover, 2001). Either the actual physical manufacturing system or its support system can be automated as long as the cost of automation is justified. If both systems are automated, a high level of integration can also be reached as in the case of computer-integrated manufacturing (CIM) systems. Groover (2001) lists examples of automation in a manufacturing system:

- Automated machines
- Transfer lines that perform multiple operations
- Automated assembly operations
- Robotic manufacturing and assembly operations
- Automated material handling and storage systems
- Automated inspection stations

Automated manufacturing systems are classified into three groups (Groover, 2001):

1. Fixed Automation Systems: Used for high production rates or volumes and very little product variety as in the case of a welding fixture used in making circular welds around pressure vessels.

2. Programmable Automation Systems: Used for batch production with small volumes and high variety as in the case of a robotic welding cell.
3. Flexible Automation Systems: Medium production rates and varieties can be covered as in the case of flexible manufacturing cells with almost no lost time for changeover from one product to another one.

Groover lists some of the reasons for automating a manufacturing entity:

1. To increase labor productivity and costs by substituting human workers with automated means
2. To mitigate skilled labor shortages as in welding and machining
3. To reduce routine and boring manual and clerical tasks
4. To improve worker safety by removing them from the point of operation in dangerous tasks such nuclear, chemical, or high energy
5. To improve product quality and repeatability
6. To reduce manufacturing lead times
7. To accomplish processes that cannot be done manually
8. To avoid the high cost of not automating

Conclusions

Manufacturing is the livelihood and future of every nation, yet it is misunderstood. It has a great role in driving engineering research and development in addition to generating wealth. Utilization of efficient and effective methods is crucial for any manufacturing enterprise to remain competitive in this global market, with intense international collaboration and rivalry. It is especially important that the industrial engineering tools used in optimizing processes and integrating those processes into systems need to be grasped to better the manufacturing processes and their systems. Automation is still a valid medium for improving the manufacturing enterprise as long as the costs of doing it are justified. Tasks too difficult to automate, life cycles that are too short, products that are very customized, and cases where demands are very unpredictably varying, cannot justify the application of automation (Hitomi, 1996).

References

Badiru, A. B. (2005), "*Product Planning and Control for Nano-Manufacturing: Application of Project Management,*" Distinguished Lecture Series, University of Central Florida, Orlando.

Badiru, A. B. (1989), "How to plan and manage manufacturing automation projects," in *Proceedings of 1989 IIE Fall Conference, Atlanta, GA, November 13–15*, 540–545.

Badiru, A. B. (1990), "Analysis of data requirements for FMS implementation is crucial to success," *Industrial Engineering*, 22(10), 29–32.

Dano, S. (1966), *Industrial Production Models: A Theoretical Study*, Springer, Vienna.

Gallager, P. (2012), *Presentation at the National Network of Manufacturing Innovation Meeting II*, Cuyahoga Community College, Cleveland, OH.

Groover, M. (2001), *Automation, Production Systems, and Computer-Integrated Manufacturing*, Prentice Hall, Upper Saddle River, NJ.

Harrington, J. Jr. (1973), *Computer Integrated Manufacturing*, Industrial Press, New York.

Hitomi, K., (1994), Moving toward manufacturing excellence for future production perspectives, *Industrial Engineering*, 26(6), 48–50.

Hitomi, K. (1996), *Manufacturing Systems Engineering: A Unified Approach to Manufacturing Technology, Production Management, and Industrial Economics*, Second Edition, Taylor & Francis, Bristol, PA.

Kalpakjian, S. and Schmid, S. R. (2006), *Manufacturing Engineering and Technology*, Fifth Edition, Pearson Prentice Hall, Upper Saddle River, NJ.

Noorani, R. (2006), *Rapid Prototyping and Applications*, John Wiley and Sons, Hoboken, NJ.

Raman, S. and Wadke, A. (2014), "Manufacturing technology," in Badiru, A. B., (Ed.), *Handbook of Industrial and Systems Engineering*, Second Edition, Taylor & Francis Group/CRC Press, Boca Raton, FL, 337–349.

Sieger, D. B. and Badiru, A. B. (1993), "An artificial neural network case study: Prediction versus classification in a manufacturing application," *Computers and Industrial Engineering*, 25(1–4), 381–384.

Sirinterlikci, A. (2000). "Thermal management and prediction of heat checking in H-13 die-casting dies," Ph.D. Dissertation, The Ohio State University, Columbus.

Sirinterlikci, A. (2014), "Manufacturing processes and systems," in Badiru, A. B. (Ed.), *Handbook of Industrial and Systems Engineering*, Second Edition, Taylor & Francis Group/CRC Press, Boca Raton, FL, 371–397.

Sirinterlikci, A., Karaman, A., Imamoglu, O., Buxton, G., Badger, P. and Dress, B. (2010), Role of biomaterials in sustainability, *2nd Annual Conference of the Sustainable Enterprises of the Future*, Pittsburgh, PA.

Sirinterlikci, A., Acors, C., Pogel, S., Wissinger, J., and Jimenez, M. (2012). "Antimicrobial technologies in design and manufacturing of medical devices," *SME Nanomanufacturing Conference*, Boston, MA.

Timms, H. L. and Pohlen, M. F. (1970), *The Production Function in Business – Decision Systems for Production and Operations Management*, Third Edition, Irwin, Homewood, IL.

U.S. Patent (1986), "Apparatus for production of three-dimensional objects by stereolithography," US 4575330 A, March 11.

chapter two

Manufacturing enterprise and technology transfer

Introduction

Why reinvent the wheel when it can be transferred and adapted from existing wheeled applications? The concepts of project management can be very helpful in planning for the adoption and implementation of new industrial technology. Due to its many interfaces, the area of industrial technology adoption and implementation is a prime candidate for the application of project planning and control techniques. Technology managers, engineers, and analysts should make an effort to take advantage of the effectiveness of project management tools. This applies to the various project management techniques that have been discussed in the preceding chapters regarding the problem of industrial technology transfer. Project management approach is presented within the context of technology adoption and implementation for industrial development. Project management guidelines are presented for industrial technology management. The Triple C model of Communication, Cooperation, and Coordination is applied as an effective tool for ensuring the acceptance of new technology. The importance of new technologies in improving product quality and operational productivity are also discussed. The chapter also outlines the strategies for project planning and control in complex technology-based operations.

Characteristics of technology transfer

To transfer technology, we must know what constitutes technology. A working definition of technology will enable us to determine how best to transfer it. A basic question that should be asked is:

What is technology?

Technology can be defined as follows:

Technology is a combination of physical and non-physical processes that make use of the latest available knowledge to achieve business, service, or production goals.

Technology is a specialized body of knowledge that can be applied to achieve a mission or purpose. The knowledge concerned could be in the form of methods, processes, techniques, tools, machines, materials, and procedures. Technology design, development, and use are driven by effective utilization of human resources and management systems. Technological progress is the result obtained when the provision of technology is used in an effective and efficient manner to improve productivity, reduce waste, improve human satisfaction, and raise the quality of life.

Technology all by itself is useless. However, when the right technology is put to the right use, with a supporting management system, it can be very effective in achieving industrialization goals. Technology implementation starts with an idea and ends with a productive industrial process. Technological progress is said to have occurred when the outputs of technology in the form of information, instrument, or knowledge that is used productively and effectively in industrial operations leads to lower production costs, better product quality, higher levels of output (from the same amount of inputs), and higher market share. The information and knowledge involved in technological progress includes those which improve the performance of management, labor, and the total resources expended for a given activity.

Technological progress plays a vital role in improving overall national productivity. Experience in developed countries such as the United States show that in the period 1870–1957, 90% of the rise in real output per man-hour can be attributed to technological progress. It is conceivable that a higher proportion of increases in per capita income is accounted for by technological change. Changes occur through improvements in the efficient use of existing technology, that is, through learning and through the adaptation of other technologies, some of which may involve different collections of technological equipment. The challenge to developing countries is how to develop the infrastructures that promote, use, adapt, and advance technological knowledge.

Most of the developing nations today face serious challenges arising not only from the world-wide imbalance of dwindling revenue from industrial products and oil, but also from major changes in a world economy that is characterized by competition, imports, and exports of not only oil, but also of basic technology, weapon systems, and electronics. If technology utilization is not given the right attention in all sectors of the national economy, the much-desired industrial development cannot occur or cannot be sustained. The ability of a nation to compete in the world market will, consequently, be stymied.

The important characteristics or attributes of a new technology may include productivity improvement, improved quality, cost savings, flexibility, reliability, and safety. An integrated evaluation must be performed to ensure that a proposed technology is justified both economically and technically. The scope and goals of the proposed technology must be established right from the beginning of the project. Table 2.1 summarizes some of the common "ability" characteristics of technology transfer for a well-rounded assessment.

An assessment of a technology transfer opportunity will entail a comparison of departmental objectives with overall organizational goals in the following areas

1. Industrial Marketing Strategy: This should identify the customers of the proposed technology. It should also address items such as the market cost of the proposed product, assessment of competition, and market share. Import and export considerations should be a key component of the marketing strategy.
2. Industry Growth and Long-range Expectations: This should address short-range expectations, long-range expectations, future competitiveness, future capability, and prevailing size and strength of the industry that will use the proposed technology.
3. National Benefit: Any prospective technology must be evaluated in terms of direct and indirect benefits to be generated by the technology. These may include product price versus value, increase in international trade, improved standard of living, cleaner environment, safer workplace, and higher productivity.
4. Economic Feasibility: An analysis of how the technology will contribute to profitability should consider past performance of the technology, incremental benefits of the new technology versus conventional technology, and value added by the new technology.
5. Capital Investment: Comprehensive economic analysis should play a significant role in the technology assessment process. This may cover an evaluation of fixed and sunk costs, cost of obsolescence, maintenance requirements, recurring costs, installation cost, space requirement cost, capital substitution options, return on investment, tax implications, cost of capital, and other concurrent projects.
6. Resource Requirements: The utilization of resources (human resources and equipment) in the pre-technology and post-technology phases of industrialization should be assessed. This may be based on material input-output flows, high value of equipment versus productivity improvement, required inputs for the technology, expected output of the technology, and utilization of technical and non-technical personnel.
7. Technology Stability: Uncertainty is a reality in technology adoption efforts. Uncertainty will need to be assessed for the initial investment, return on investment, payback period, public reactions, environmental impact, and volatility of the technology.

Table 2.1 The "ability" of manufacturing technology

Characteristics	Definitions, questions, and implications
Adaptability	Can the technology be adapted to fit the needs of the organization? Can the organization adapt to the requirements of the technology?
Affordability	Can the organization afford the technology in terms of first-cost, installation cost, sustainment cost, and other incidentals?
Capability	What are the capabilities of the technology with respect to what the organization needs? Can the technology meet the current and emerging needs of the organization?
Compatibility	Is the technology compatible with existing software and hardware?
Configurability	Can the technology be configured for the existing physical infrastructure available within the organization?
Dependability	Is the technology dependable enough to produce the outputs expected?
Desirability	Is the particular technology desirable for the prevailing operating environment of the organization? Are there environmental issues and/or social concerns related to the technology?
Expandability	Can the technology be expanded to fit the changing needs of the organization?
Flexibility	Does the technology have flexible characteristics to accomplish alternate production requirements?
Interchangeability	Can the technology be interchanged with currently available tools and equipment in the organization? In case of operational problems, can the technology be interchanged with something else?
Maintainability	Does the organization have the wherewithal to maintain the technology?
Manageability	Does the organization have adequate management infrastructure to acquire and use the technology?
Re-configurability	When operating conditions change or organizational infrastructure changes, can the technology be re-configured to meet new needs?
Reliability	Is the technology reliable in terms of technical, physical, and/or scientific characteristics?
Stability	Is the technology mature and stable enough to warrant an investment within the current operating scenario?
Sustainability	Is the organization committed enough to sustain the technology for the long haul? Is the design of the technology sound and proven to be sustainable?
Volatility	Is the technology devoid of volatile developments? Is the source of the technology devoid of political upheavals and/or social unrests?

8. National Productivity Improvement: An analysis of how the technology may contribute to national productivity may be verified by studying industrial throughput, efficiency of production processes, utilization of raw materials, equipment maintenance, absenteeism, learning rate, and design-to-production cycle.

Emergence of new technology

New industrial and service technologies have been gaining more attention in recent years. This is due to the high rate at which new productivity improvement technologies are being developed. The fast pace of new technologies has created difficult implementation and management problems for many organizations. New technology can be successfully implemented only if it is viewed as a system whose various components are evaluated within an integrated managerial framework. Such a framework is provided by a project management approach. A multitude of new technologies has emerged in recent years. It is important to consider the peculiar characteristics of a new technology before establishing adoption and implementation strategies. The justification for the adoption of a new technology is usually a combination of several factors rather than a single characteristic of the technology. The potential of a specific technology to contribute to industrial development goals must be carefully assessed. The technology assessment process should explicitly address the following questions:

What is expected from the new technology?
Where and when will the new technology be used?
How is the new technology similar to or different from existing technologies?
What is the availability of technical personnel to support the new technology?
What administrative support is needed for the new technology?
Who will use the new technology?
How will the new technology be used?
Why is the technology needed?

The development, transfer, adoption, utilization, and management of technology is a problem that is faced in one form or another by business, industry, and government establishments. Some of the specific problems in technology transfer and management include the following:

• Controlling technological change
• Integrating technology objectives
• Shortening the technology transfer time

- Identifying a suitable target for technology transfer
- Coordinating the research and implementation interface
- Formal assessment of current and proposed technologies
- Developing accurate performance measures for technology
- Determining the scope or boundary of technology transfer
- Managing the process of entering or exiting a technology
- Understanding the specific capability of a chosen technology
- Estimating the risk and capital requirements of a technology

Integrated managerial efforts should be directed at the solution of the problems stated above. A managerial revolution is needed in order to cope with the ongoing technological revolution. The revolution can be initiated by modernizing the longstanding and obsolete management culture relating to technology transfer. Some of the managerial functions that will need to be addressed when developing a technology transfer strategy include the following:

1. Development of a technology transfer plan.
2. Assessment of technological risk.
3. Assignment/reassignment of personnel to implement the technology transfer.
4. Establishment of a transfer manager and a technology transfer office. In many cases, transfer failures occur because no individual has been given the responsibility to ensure the success of technology transfer.
5. Identification and allocation of the resources required for technology transfer.
6. Setting of guidelines for technology transfer. For example,
   a. Specification of phases (development, testing, transfer, etc.)
   b. Specification of requirements for inter-phase coordination
   c. Identification of training requirements
   d. Establishment and implementation of performance measurement.
7. Identify key factors (both qualitative and quantitative) associated with technology transfer and management.
8. Investigate how the factors interact and develop the hierarchy of importance for the factors.
9. Formulate a loop system model that considers the forward and backward chains of actions needed to effectively transfer and manage a given technology.
10. Track the outcome of the technology transfer.

Technological developments in many industries appear in scattered, narrow, and isolated areas within a few selected fields. This makes technology efforts to be rarely coordinated, thereby, hampering the benefits of technology. The optimization of technology utilization is, thus, very difficult. To overcome this problem and establish the basis for effective technology transfer and management, an integrated approach must be followed. An integrated approach will be applicable to technology transfer between any two organizations whether public or private.

Some nations concentrate on the acquisition of bigger, better, and faster technology. But little attention is given to how to manage and coordinate the operations of the technology once it arrives. When technology fails, it is not necessarily because the technology is deficient. Rather, it is often the communication, cooperation, and coordination functions of technology management that are deficient. Technology encompasses factors and attributes beyond mere hardware, software, and "skinware," which refers to people issues affecting the utilization of technology. This may involve social-economic and cultural issues of using certain technologies. Consequently, technology transfer involves more than the physical transfer of hardware and software. Several flaws exist in the common practices of technology transfer and management. These flaws include in the following:

- Poor fit: This relates to an inadequate assessment of the need of the organization receiving the technology. The target of the transfer may not have the capability to properly absorb the technology.
- Premature transfer of technology: This is particularly acute for emerging technologies that are prone to frequent developmental changes.
- Lack of focus: In the attempt to get a bigger share of the market or gain an early lead in the technological race, organizations frequently force technology in many incompatible directions.
- Intractable implementation problems: Once a new technology is in place, it may be difficult to locate sources of problems that have their roots in the technology transfer phase itself.
- Lack of transfer precedents: Very few precedents are available on the management of brand new technology. Managers are, thus, often unprepared for their new technology management responsibilities.
- Stuck on technology: Unworkable technologies sometimes continue to be recycled needlessly in the attempt to find the "right" usage.
- Lack of foresight: Due to the nonexistence of a technology transfer model, managers may not have a basis against which they can evaluate future expectations.
- Insensitivity to external events: Some external events that may affect the success of technology transfer may include trade barriers, taxes, and political changes.
- Improper allocation of resources: There are usually not enough resources available to allocate to technology alternatives. Thus, a technology transfer priority must be developed.

The following steps provide a specific guideline for pursuing the implementation of manufacturing technology transfer:

1. Find a suitable application.
2. Commit to an appropriate technology.
3. Perform economic justification.
4. Secure management support for the chosen technology.
5. Design the technology implementation to be compatible with existing operations.
6. Formulate project management approach to be used.
7. Prepare the receiving organization for the technology change.
8. Install the technology.
9. Maintain the technology.
10. Periodically review the performance of the technology based on prevailing goals.

Technology transfer modes

The transfer of technology can be achieved in various forms. Project management provides an effective means of ensuring proper transfer of technology. Three technology transfer modes are presented here to illustrate basic strategies for getting one technological product from one point (technology source) to another point (technology target). A conceptual integrated model of the interaction between the technology source and target is presented in Figure 2.1.

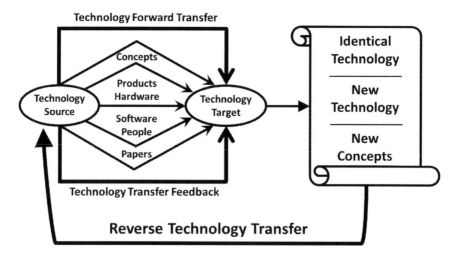

Figure 2.1 Technology transfer modes.

The university-industry interaction model presented in this book can be used as an effective mechanism for facilitating technology transfer. Industrial technology application centers may be established to serve as a unified point for linking technology sources with interested targets. The center will facilitate interactions between business establishments, academic institutions, and government agencies to identify important technology needs. With reference to Figure 2.1, technology can be transferred in one or a combination of the following strategies:

1. Transfer of complete technological products: In this case, a fully developed product is transferred from a source to a target. Very little product development effort is carried out at the receiving point. However, information about the operations of the product is fed back to the source so that necessary product enhancements can be pursued. So, the technology recipient generates product information which facilitates further improvement at the technology source. This is the easiest mode of technology transfer and the most tempting. Developing nations are particularly prone to this type of transfer. Care must be exercised to ensure that this type of technology transfer does not degenerate into "machine transfer." It should be recognized that machines alone do not constitute technology.

2. Transfer of technology procedures and guidelines: In this technology transfer mode, procedures (e.g., blueprints) and guidelines are transferred from a source to a target. The technology blueprints are implemented locally to generate the desired services and products. The use of local raw materials and manpower is encouraged for the local production. Under this mode, the implementation of the transferred technology procedures can generate new operating procedures that can be fed back to enhance the original technology. With this symbiotic arrangement, a loop system is created whereby both the transferring and the receiving organizations derive useful benefits.

3. Transfer of technology concepts, theories, and ideas: This strategy involves the transfer of the basic concepts, theories, and ideas behind a given technology. The transferred elements can then be enhanced, modified, or customized within local constraints to generate new technological products. The local modifications and enhancements have the potential to generate an identical technology, a new related technology, or a new set of technology concepts, theories, and ideas. These derived products may then be transferred back to the original technology source as new technological enhancements. Figure 2.2 presents a specific cycle for local adaptation and modification of technology. An academic institution is a good potential source for the transfer of technology concepts, theories, and ideas.

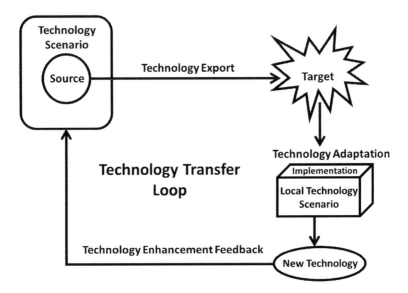

Figure 2.2 Local adaptation and enhancement of technology.

It is very important to determine the mode in which technology will be transferred for manufacturing purposes. There must be a concerted effort by people to make the transferred technology work within local infrastructure and constraints. Local innovation, patriotism, dedication, and willingness to adapt technology will be required to make technology transfer successful. It will be difficult for a nation to achieve industrial development through total dependence on transplanted technology. Local adaptation will always be necessary.

Technology change-over strategies

Any development project will require changing from one form of technology to another. The implementation of a new technology to replace an existing (or a nonexistent) technology can be approached through one of several options. Some options are more suitable than others for certain types of technologies. The most commonly used technology change-over strategies include the following:

Parallel Change-over: In this case, the existing technology and the new technology operate concurrently until there is confidence that the new technology is satisfactory.

Direct Change-over: In this approach, the old technology is removed totally and the new technology takes over. This method is recommended only when there is no existing technology or when both technologies cannot be kept operational due to incompatibility or cost considerations.

Phased Change-over: In this incremental change-over method, modules of the new technology are gradually introduced one at a time using either direct or parallel change-over.

Pilot Change-over: In this case, the new technology is fully implemented on a pilot basis in a selected department within the organization.

Post-implementation evaluation

The new technology should be evaluated only after it has reached a steady-state performance level. This helps to avoid the bias that may be present at the transient stage due to personnel anxiety, lack of experience, or resistance to change. The system should be evaluated for the following aspects:

- Sensitivity to data errors
- Quality and productivity
- Utilization level
- Response time
- Effectiveness

Technology systems integration

With the increasing shortages of resources, more emphasis should be placed on the sharing of resources. Technology resource sharing can involve physical equipment, facilities, technical information, ideas, and related items. The integration of technologies facilitates the sharing of resources. Technology integration is a major effort in technology adoption and implementation. Technology integration is required for proper product coordination. Integration facilitates the coordination of diverse technical and managerial efforts to enhance organizational functions, reduce cost, improve productivity, and increase the utilization of resources. Technology integration ensures that all performance goals are satisfied with a minimum of expenditure of time and resources. It may require the adjustment of functions to permit sharing of resources, development of new policies to accommodate product integration, or realignment of managerial responsibilities. It can affect both hardware and software components of an organization. Important factors in technology integration include the following:

- Unique characteristics of each component in the integrated technologies
- Relative priorities of each component in the integrated technologies

- How the components complement one another
- Physical and data interfaces between the components
- Internal and external factors that may influence the integrated technologies
- How the performance of the integrated system will be measured

Role of government in technology transfer

The malignant policies and operating characteristics of some of the governments in underdeveloped countries have contributed to stunted growth of technology in those parts of the world. The governments in most developing countries control the industrial and public sectors of the economy. Either people work for the government or serve as agents or contractors for the government. The few industrial firms that are privately owned depend on government contracts to survive. Consequently, the nature of the government can directly determine the nature of industrial technological progress.

The operating characteristics of most of the governments perpetuate inefficiency, corruption, and bureaucratic bungles. This has led to a decline in labor and capital productivity in the industrial sectors. Using the Pareto distribution, it can be estimated that in most government-operated companies, there are eight administrative workers for every two production workers. This creates a non-productive environment that is skewed toward hyper-bureaucracy. The government of a nation pursuing industrial development must formulate and maintain an economic stabilization policy. The objective should be to minimize the sacrifice of economic growth in the short run and while maximizing long-term economic growth. To support industrial technology transfer efforts, it is essential that a conducive national policy be developed.

More emphasis should be placed on industry diversification, training of the workforce, supporting financial structure for emerging firms and implementing policies that encourage productivity in a competitive economic environment. Appropriate foreign exchange allocation, tax exemptions, bank loans for emerging businesses, and government-guaranteed low interest loans for potential industrial entrepreneurs are some of the favorable policies to spur growth and development of the industrial sector.

Improper trade and domestic policies have adversely affected industrialization in many countries. Excessive regulations that cause bottlenecks in industrial enterprises are not uncommon. The regulations can take the form of licensing, safety requirements, manufacturing value-added quota requirements, capital contribution by multinational firms, and high domestic production protection. Although regulations are needed for industrial operations, excessive controls lead to low returns from the industrial sectors. For example, stringent regulations on foreign exchange allocation and control have led to the closure of industrial plants in some countries. The firms that cannot acquire essential raw materials, commodities, tools, equipment, and new technology from abroad due to foreign exchange restrictions are forced to close and lay off workers.

Price controls for commodities are used very often by developing countries especially when inflation rates for essential items are high. The disadvantages involved in price control of industrial goods include restrictions of the free competitive power of available goods in relation to demand and supply, encouragement of inefficiency, promotion of dual markets, distortion of cost relationships, and increase in administrative costs involved in producing goods and services.

US templates for technology transfer

One way that a government can help facilitate industrial technology transfer involves the establishment of technology transfer centers within appropriate government agencies. A good example of this approach can be seen in the government-sponsored technology transfer program by the US National Aeronautics and Space Administration (NASA). In the Space Act of 1958, the US Congress charged NASA with a responsibility to provide for the widest practical and appropriate dissemination of information concerning its activities and the results achieved from those activities. With this technology transfer responsibility, technology developed in the United States' space program is available for use by the nation's business and industry.

In order to accomplish technology transfer to industry, NASA established a Technology Utilization Program (TUP) in 1962. The technology utilization program uses several avenues to disseminate information on NASA technology. The avenues include the following:

- Complete, clear, and practical documentation is required for new technology developed by NASA and its contractors. These are available to industry through several publications produced by NASA. An example is a monthly, *Tech Briefs*, which outlines technology innovations. This is a source of prompt technology information for industry.
- Industrial Application Centers (IACs) were developed to serve as repositories for vast computerized data on technical knowledge. The IACs are located at academic institutions around the country. All the centers have access to a large database containing millions of NASA documents. With this database, industry can have access to the latest technological information quickly. The funding for the centers is obtained through joint contributions from several sources including

NASA, the sponsoring institutions, and state government subsidies. Thus, the centers can provide their services at very reasonable rates.

- NASA operates a Computer Software Management and Information Center (COSMIC) to disseminate computer programs developed through NASA projects. COSMIC, which is located at the University of Georgia, has a library of thousands of computer programs. The center publishes an annual index of available software.

In addition to the specific mechanisms discussed above, NASA undertakes application engineering projects. Through these projects, NASA collaborates with industry to modify aerospace technology for use in industrial applications. To manage the application projects, NASA established a Technology Application Team (TAT), consisting of scientists and engineers from several disciplines. The team interacts with NASA field centers, industry, universities, and government agencies. The major mission of the team interactions is to define important technology needs and identify possible solutions within NASA. NASA application engineering projects are usually developed in a five-phase approach with go or no-go decisions made by NASA and industry at the completion of each phase. The five phases are outlined below:

1. NASA and the Technology Applications Team meet with industry associations, manufacturers, university researchers, and public sector agencies to identify important technology problems that might be solved by aerospace technology.
2. After a problem is selected, it is documented and distributed to the Technology Utilization Officer at each of NASA's field centers. The officer in turn distributes the description of the problem to the appropriate scientists and engineers at the center. Potential solutions are forwarded to the team for review. The solutions are then screened by the problem originator to assess the chances for technical and commercial success.
3. The development of partnerships and a project plan to pursue the implementation of the proposed solution. NASA joins forces with private companies and other organizations to develop an applications engineering project. Industry participation is encouraged through a variety of mechanisms such as simple letters of agreement or joint endeavor contracts. The financial and technical responsibilities of each organization are specified and agreed upon.
4. At this point, NASA's primary role is to provide technical assistance to facilitate utilization of the technology. The costs for these projects are usually shared by NASA and the participating companies. The proprietary information provided by the companies and their rights to new discoveries are protected by NASA.

5. The final phase involves the commercialization of the product. With the success of commercialization, the project would have widespread impact. Usually, the final product development, field testing, and marketing are managed by private companies without further involvement from NASA.

Through this well-coordinated government-sponsored technology transfer program, NASA has made significant contributions to the US industry. The results of NASA's technology transfer abound in numerous consumer products either in subtle forms or in clearly identifiable forms. Food preservation techniques constitute one area of NASA's technology transfer that has had a significant positive impact on the society. Although the specific organization and operation of the NASA technology transfer programs have changed in name or in deed over the years, the basic descriptions outlined above remain a viable template for how to facilitate manufacturing technology transfer. Other nations can learn from NASA's technology transfer approach. In a similar government-backed strategy, the US Air Force Research Lab (AFRL) also has very structured programs for transferring non-classified technology to the industrial sector.

The major problem in developing nations is not the lack of good examples to follow. Rather, the problem involves not being able to successfully manage and sustain a program that has proven successful in other nations. It is believed that a project management approach can help in facilitating success with manufacturing technology transfer efforts.

Pathway to national strategy

Most of the developing nations depend on technologies transferred from developed nations to support their industrial base. This is partly due to a lack of local research and development programs, development funds, and work force needed to support such activity. Advanced technology is desired by most industries in developing countries because of its potential to increase output. The adaptability of advanced technology to industries in a developing country is a complex and difficult task. Evidence in most manufacturing firms that operate in developing countries reveal that advanced technology can lead to machine down time because the local plants do not have the maintenance and repair facilities to support the use of advanced technology.

In some situations, most firms cannot afford the high cost of maintenance associated with the use of foreign technology. One way to solve the transfer of technology problem is by establishing local design centers for developing nation's industrial sectors, such centers will design and adapt technology for local usage. In addition, such centers will also work on adapting full assembled machinery from developed countries. However, the fertile ground for the introduction of appropriate technology is where people are already organized under a good system of government, production, marketing, and continuing improvement in standard of living. Developing countries must place more emphasis on the production of useful, consumable goods and services. One useful strategy to ensure a successful transfer of technology is by providing training services that will ensure proper repair and maintenance of technology hardware. It is important that a nation trying to transfer technology should have access to a broad-based body of technical information and experience. A plan of technical information sharing between suppliers and users must be assured. The transfer of technology also requires a reliable liaison between the people who develop the ideas, their agents, and the people who originate the concepts. Technology transfer is only complete when the technology becomes generally accepted in the workplace. Local efforts are needed in tailoring technological solutions to local problems. Technicians and engineers must be trained to assume the role of technology custodians so that implementation and maintenance problems are minimized. A strategy for minimizing the technology transfer disconnection is to set up central repair shops dedicated to making spare parts and repairing equipment on a timely basis to reduce industrial machine down time. If the utilization level of equipment is increased, there will be an increase in the productive capacity of the manufacturer. Improving maintenance and repair centers in developing countries will provide an effective way of assisting emerging firms in developing countries where dependence on transferred technology is prevalent. There should also be a strategy to develop appropriate local technology to support the goals of industrialization. This is important because transferred technology may not be fully suitable or compatible with local product specifications. For example, many nations have experienced the failure of transferred food processing technology because the technology was not responsive to the local diets, ingredients, and food preparation practices. One way to accomplish the development of local technology is to encourage joint research efforts between academic institutions and industrial firms. Some of the chapters in this book explicitly address university-industry collaborations. The design centers suggested earlier can help in this process. In addition to developing new local technologies, existing technologies should be calibrated for local usage and the higher production level required for industrialization.

The government of developing nations must assume leadership roles in encouraging research and development activities, awarding research grants to universities and private organizations geared toward seeking better ways for developing and adapting technologies for local usage. Effective innovations and productivity improvement cannot happen without adequate public and private sector policies. A nation that does not have an effective policy for productivity management and technology advancement will always find itself in a cycle of unstable economy and business crisis. Increases in real product capital, income level, and quality of life are desirable goals that are achievable through effective policies that are executed properly. The following recommendations are offered to encourage industrial growth and technological progress:

1. Encourage a free enterprise system that believes in and practices fair competition. Discourage protectionism and remove barriers to allow free trade.
2. Avoid nationalization of assets of companies jointly developed by citizens of developing countries and multinationals. Encourage joint industrial ventures among nations.
3. Both public and private sectors of the economy should encourage and invest in improving national education standards for citizens at various levels.
4. Refrain from dependence on borrowed money and subsidy programs. Create productive enterprises locally that provide essential commodities for local consumptions and exports.
5. Both public and private sectors should invest more on systems and programs, research, and development that generate new breakthroughs in technology and methods for producing food rather than war instruments.
6. The public sector should establish science and technology centers to foster the development of new local technology, productivity management techniques, and production methodologies.
7. Encourage strong partnership between government, industry, and academic communities in formulating and executing national development programs.
8. Governments and financial institutions should provide low interest loans to entrepreneurs willing to take the risk in producing essential goods and services through small-scale industries.
9. Implement a tax structure that is equitable and one that provides incentives for individuals and businesses that are working to expand employment opportunities and increase the final output of the national economy.
10. Refrain from government control of productive enterprises. Such controls only create grounds for fraud and corruption. Excessive regulations should be discouraged.
11. Periodically assess the ratio of administrative workers to production workers, administrative workers to service workers, in both private and public sectors. Implement actions to reduce excessive administrative procedures and bureaucratic bottlenecks that impede productivity and technological progress.

12. Encourage organizations and firms to develop and implement strategies, methods, and techniques in a framework of competitive and long-term performance.
13. Trade policy laws and regulations should be developed and enforced in a framework that recognizes fair competition in a global economy.
14. Create a national productivity, science, and technology council to facilitate the implementation of good programs, enhance cooperation between private and public sectors of the economy, redirect the economy toward growth strategies, and encourage education and training of the work force.
15. Implement actions that insure stable fiscal, monetary, and income policies. Refrain from wage and price control by political means. Let the elements of the free enterprise system control inflation rate, wages, and income distribution.
16. Encourage morale standards that take pride in excellence, work ethics, and a value system that encourages pride in consumer products produced locally.
17. Encourage individuals and businesses to protect full employment programs, maintain income levels by investing in local ventures rather than exporting capital abroad.
18. Both the public and private sectors of the economy should encourage and invest in re-training of the workforce as new technology and techniques are introduced for productive activities.
19. Make use of the expertise of nations that are professionally based abroad. This is an excellent source of experience for local technology development.
20. Arrange for annual conferences, seminars, and workshops to exchange ideas between researchers, entrepreneurs, practitioners, and managers with the focus on the processes required for industrial development.

Using PICK chart for technology selection

The question of which technology is appropriate to transfer in or transfer out is relevant for technology transfer considerations. While several methods of technology selection are available, this book recommends methods that combine qualitative and quantitative factors. The analytical hierarchy process (AHP) is one such method. Another useful, but less publicized method is the PICK chart. The PICK chart was originally developed by the Lockheed Martin Corporation to identify and prioritize improvement opportunities in the company's process improvement applications. The technique is just one of the several decision tools available in process improvement endeavors. It is a very effective technology selection tool used to categorize ideas and opportunities. The purpose is to qualitatively help identify the most useful ideas. A 2×2 grid is normally drawn on a white board or large flip-chart. Ideas that were written on sticky notes by team members are placed on the grid based on a group assessment of the payoff relative the level of difficulty. The PICK acronym comes from the labels for each of the quadrants of the grid: **P**ossible (easy, low payoff), **I**mplement (easy, high payoff), **C**hallenge (hard, high payoff), and **K**ill (hard, low payoff). The PICK chart quadrants are summarized as follows:

Possible (easy, low payoff)	→ Third quadrant
Implement (easy, high payoff)	→ Second quadrant
Challenge (hard, high payoff)	→ First quadrant
Kill (hard, low payoff).	→ Fourth quadrant

The primary purpose is to help identify the most useful ideas, especially those that can be accomplished immediately with little difficulty. These are called "Just-Do-Its." The general layout of the PICK chart grid is shown in Figure 2.3. The PICK process is normally done subjectively by a team of decision-makers under a group decision process. This can lead to bias and protracted debate of where each item belongs. It is desired to improve the efficacy of the process by introducing some quantitative analysis. Badiru and Thomas (2013) present a methodology to achieve a quantification of the PICK selection process. The PICK chart is often criticized for its subjective rankings and lack of quantitative analysis. The approach presented by Badiru and Thomas (2013) alleviates such concerns by normalizing and quantifying the process of integrating the subjective rakings by those involved in the group PICK process. Human decision

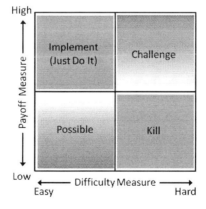

Figure 2.3 Basic layout of the PICK chart.

is inherently subjective. All we can do is to develop techniques to mollify the subjective inputs rather than compounding them with subjective summarization.

PICK chart quantification methodology

The placement of items into one of the four categories in a PICK chart is done through expert ratings, which are often subjective and non-quantitative. In order to put some quantitative basis to the PICK chart analysis, Badiru and Thomas (2013) present the methodology of dual numeric scaling on the impact and difficulty axes. Suppose each technology is ranked on a scale of one to ten and plotted accordingly on the PICK chart. Then, each project can be evaluated on a binomial pairing of the respective rating on each scale. Note that a high rating along the x axis is desirable while a high rating along the y axis is not desirable. Thus, a composite rating involving x and y must account for the adverse effect of high values of y. A simple approach is to define $y' = (11-y)$, which is then used in the composite evaluation. If there are more factors involved in the overall project selection scenario, the other factors can take on their own lettered labeling (e.g., a, b, c, z, etc.). Then, each project will have an n-factor assessment vector. In its simplest form, this approach will generate a rating such as the following:

$$PICK_{R,i}(x, y') = x + y'$$

where:

$PICK_{R,i}(x,y) = $ PICK rating of project i ($i = 1, 2, 3, \ldots, n$)

$n = $ number of projects under consideration

$x = $ rating along the impact axis ($1 \leq x \leq 10$)

$y = $ rating along the difficulty axis ($1 \leq y \leq 10$)

$y' = (11-y)$

If $x + y'$ is the evaluative basis, then each technology's composite rating will range from 2 to 20, 2 being the minimum and 20 being the maximum possible. If $(x)(y)$ is the evaluative basis, then each project's composite rating will range from 1 to 100. In general, any desired functional form may be adopted for the composite evaluation. Another possible functional form is:

$$PICK_{R,i}(x, y'') = f(x, y'')$$

$$= (x + y'')^2,$$

where y'' is defined as needed to account for the converse impact of the axes of difficulty. The above methodology provides a quantitative measure for translating the entries in a conventional PICK chart into an analytical technique to rank the technology alternatives, thereby reducing the level of subjectivity in the final decision. The methodology can be extended to cover cases where a technology has the potential to create negative impacts, which may impede organizational advancement.

The quantification approach facilitates a more rigorous analytical technique compared to traditional subjective approaches. One concern is that although quantifying the placement of alternatives on the PICK chart may improve the granularity of relative locations on the chart, it still does not eliminate the subjectivity of how the alternatives are assigned to quadrants in the first place. This is a recognized feature of many decision tools. This can be mitigated by the use of additional techniques that aid decision-makers to refine their choices. The analytic hierarchy process (AHP) could be useful for this purpose. Quantifying subjectivity is a continuing challenge in decision analysis. The PICK chart quantification methodology offers an improvement over the conventional approach.

Although the PICK chart has been used extensively in industry, there are few published examples in the open literature. The quantification approach presented by Badiru and Thomas (2013) may expand interest and applications of the PICK chart among technology researchers and practitioners. The steps for implementing a PICK chart are summarized below:

Step 1: On a chart, place the subject question. The question needs to be asked and answered by the team at different stages to be sure that the data that is collected is relevant.

Step 2: Put each component of the data on a different note like a sticky note or small cards. These notes should be arranged on the left side of the chart.

Step 3: Each team member must read all notes individually and consider its importance. The team member should decide whether the element should or should not remain a fraction of the significant sample. The notes are then removed and moved to the other side of the chart. Now, the data is condensed enough to be processed for a particular purpose by means of tools that allow groups to reach a consensus on priorities of subjective and qualitative data.

Step 4: Apply the quantification methodology presented above to normalize the qualitative inputs of the team.

DEJI model for technology integration

Technology is at the intersection of efficiency, effectiveness, and productivity. Efficiency provides the framework for quality in terms of resources and inputs required to achieve the desired level of quality. Effectiveness comes into play with respect to the application of product quality to meet specific needs and requirements of an organization. Productivity is an essential factor in the pursuit of quality as it relates to the throughput of a production system. To achieve the desired levels of quality, efficiency, effectiveness, and productivity, a new technology integration framework must be adopted. This section presents a technology integration model for design, evaluation, justification, and integration (DEJI) based on the product development application presented by Badiru (2012). The model is relevant for research and development efforts in industrial development and technology applications. The DEJI model encourages the practice of building quality into a product right from the beginning so that the product or technology integration stage can be more successful. The essence of the model is summarized in Table 2.2.

Design for technology implementation

The design of quality in product development should be structured to follow point-to-point transformations. A good technique to accomplish this is the use of state-space transformation, with which we can track the evolution of a product from the concept stage to a final product stage. For the purpose of product quality design, the following definitions are applicable:

Product state: A state is a set of conditions that describe the product at a specified point in time. The *state* of a product refers to a performance characteristic of the product which relates input to output such that a knowledge of the input function over time and the state of the product at time $t = t_0$ determines the expected output for $t \geq t_0$. This is particularly important for assessing where the product

Table 2.2 DEJI model for technology integration

DEJI model	Characteristics	Tools & techniques
Design	Define goals	Parametric assessment
	Set performance metrics	Project state transition
	Identify milestones	Value stream analysis
Evaluate	Measure parameters	Pareto distribution
	Assess attributes	Life cycle analysis
	Benchmark results	Risk assessment
Justify	Assess economics	Benefit-cost ratio
	Assess technical output	Payback period
	Align with goals	Present value
Integrate	Embed in normal operation	SMART concept
	Verify symbiosis	Process improvement
	Leverage synergy	Quality control

stands in the context of new technological developments and the prevailing operating environment.

Product state-space: A product *state-space* is the set of all possible states of the product lifecycle. State-space representation can solve product design problems by moving from an initial state to another state, and eventually to the desired end-goal state. The movement from state to state is achieved by means of actions. A goal is a description of an intended state that has not yet been achieved. The process of solving a product problem involves finding a sequence of actions that represents a solution path from the initial state to the goal state. A state-space model consists of state variables that describe the prevailing condition of the product. The state variables are related to inputs by mathematical relationships. Examples of potential product state variables include schedule, output quality, cost, due date, resource, resource utilization, operational efficiency, productivity throughput, and technology alignment. For a product described by a system of components, the state-space representation can follow the quantitative metric below:

$$Z = f(z, x); Y = g(z, x)$$

where f and g are vector-valued functions. The variable Y is the output vector while the variable x denotes the inputs. The state vector Z is an intermediate vector relating x to y. In generic terms, a product is transformed from one state to another by a driving function that produces a transitional relationship given by:

$$S_s = f(x \mid S_p) + e,$$

where:
    S_s = subsequent state
    x = state variable
    S_p = the preceding state
    e = error component.

The function f is composed of a given action (or a set of actions) applied to the product. Each intermediate state may represent a significant milestone in the project. Thus, a descriptive state-space model facilitates an analysis of what actions to apply in order to achieve the next desired product state. The state-space representation can be expanded to cover several components within the technology integration framework. Hierarchical linking of product elements provides an expanded transformation structure. The product state can be expanded in accordance with implicit requirements. These requirements might include grouping of design elements, linking precedence requirements (both technical and procedural), adapting to new technology developments, following required communication links, and accomplishing reporting requirements. The actions to be taken at each state depend on the prevailing product conditions. The nature of subsequent alternate states depends on what actions are implemented. Sometimes there are multiple paths that can lead to the desired end result. At other times, there exists only one unique path to the desired objective. In conventional practice, the characteristics of the future states can only be recognized after the fact, thus, making it impossible to develop adaptive plans. In the implementation of the DEJI model, adaptive plans can be achieved because the events occurring within and outside the product state boundaries can be considered. If we describe a product by P state variables s_i, then the composite state of the product at any given time can be represented by a vector $\mathbf{S}$ containing P elements. That is,

$$\mathbf{S} = \{s_1, s_2, \ldots, s_P\}$$

The components of the state vector could represent either quantitative or qualitative variables (e.g., cost, energy, color, time). We can visualize every state vector as a point in the state space of the product. The representation is unique since every state vector corresponds to one and only one point in the state-space. Suppose we have a set of actions (transformation agents) that we can apply to the product information so as to change it from one state to another within the project state-space. The transformation will change a state vector into another state vector. A transformation may be a change in raw material or a change in design approach. The number of transformations available for a product characteristic may be finite or unlimited. We can construct trajectories that describe the potential states of a product evolution as we apply successive transformations with respect to technology forecasts. Each transformation may be repeated as many times as needed. Given an initial state $\mathbf{S}_0$, the sequence of state vectors is represented by the following:

$$\mathbf{S}_n = T_n(\mathbf{S}_{n-1}).$$

The state-by-state transformations are then represented as $\mathbf{S}_1 = T_1(\mathbf{S}_0)$; $\mathbf{S}_2 = T_2(\mathbf{S}_1)$; $\mathbf{S}_3 = T_3(\mathbf{S}_2)$; ...; $\mathbf{S}_n = T_n(\mathbf{S}_{n-1})$. The final state, $\mathbf{S}_n$, depends on the initial state $\mathbf{S}$ and the effects of the actions applied.

Evaluation of technology

A product can be evaluated on the basis of cost, quality, schedule, and meeting requirements. There are many quantitative metrics that can be used in evaluating a product at this stage. Learning curve productivity is one relevant technique that can be used because it offers an evaluation basis of a product with respect to the concept of growth and decay. The half-life extension (Badiru, 2012) of the basic learning is directly applicable because the half-life of the technologies going into a product can be considered. In today's technology-based operations, retention of learning may be threatened by fast-paced shifts in operating requirements. Thus, it is of interest to evaluate the half-life properties of new technologies as the impact the overall product quality. Information about the half-life can tell us something about the sustainability of learning-induced technology performance. This is particularly useful for designing products whose life cycles stretch into the future in a high-tech environment.

Justification of technology

We need to justify a program on the basis of quantitative value assessment. The Systems Value Model (SVM) is a good quantitative technique that can be used here for project justification on the basis of value. The model provides a heuristic decision aid for comparing project alternatives. It is presented here again for the present context. Value is represented as a deterministic vector function that indicates the value of tangible and intangible attributes that characterize the project. It is represented as $V = f\left(A_1, A_2, \ldots, A_p\right)$, where V is the assessed value and the A values are quantitative measures or attributes. Examples of product attributes are quality, throughput, manufacturability, capability, modularity, reliability, interchangeability, efficiency, and cost performance. Attributes are considered to be a combined function of factors. Examples of product factors are market share, flexibility, user acceptance, capacity utilization, safety, and design functionality. Factors are themselves considered to be composed of indicators. Examples of indicators are debt ratio, acquisition volume, product responsiveness, substitutability, lead time, learning curve, and scrap volume. By combining the above definitions, a composite measure of the operational value of a product can be quantitatively assessed. In addition to the quantifiable factors, attributes, and indicators that impinge upon overall project value, the human-based subtle factors should also be included in assessing overall project value.

Integration of technology

Without being integrated, a system will be in isolation and it may be worthless. We must integrate all the elements of a system on the basis of alignment of functional goals. The overlap of systems for integration purposes can conceptually be viewed as projection integrals by considering areas bounded by the common elements of sub-systems. Quantitative metrics can be applied at this stage for effective assessment of the technology state. Trade-off analysis is essential in technology integration. Pertinent questions include the following:

What level of trade-offs on the level of technology are tolerable?
What is the incremental cost of more technology?
What is the marginal value of more technology?
What is the adverse impact of a decrease in technology utilization?
What is the integration of technology over time? In this respect, an integral of the form below may be suitable for further research:

$$I = \int_{t_1}^{t_2} f(q)dq,$$

where:
I = integrated value of quality
$f(q)$ = functional definition of quality
t_1 = initial time
t_2 = final time within the planning horizon.

Presented below are guidelines and important questions relevant for technology integration.

- What are the unique characteristics of each component in the integrated system?
- How do the characteristics complement one another?
- What physical interfaces exist among the components?
- What data/information interfaces exist among the components?
- What ideological differences exist among the components?
- What are the data flow requirements for the components?
- What internal and external factors are expected to influence the integrated system?
- What are the relative priorities assigned to each component of the integrated system?
- What are the strengths and weaknesses of the integrated system?
- What resources are needed to keep the integrated system operating satisfactorily?
- Which organizational unit has primary responsibility for the integrated system?

The recommended approach of the DEJI model will facilitate a better alignment of product technology with future development and needs. The stages of the model require research for each new product with respect to design, evaluation, justification, and integration. Existing analytical tools and techniques can be used at each stage of the model.

Conclusion

Technology transfer is a great avenue to advancing industrialization. This chapter has presented a variety of principles, tools, techniques, and strategies useful for managing technology transfer. Of particular emphasis in the chapter is the management aspects of technology transfer. The technical characteristics of the technology of interest are often well understood. What is often lacking is an appreciation of the technology management requirements for achieving a successful technology transfer. This chapter presents the management aspects of manufacturing technology transfer.

References

Badiru, A. B. (2012), "Application of the DEJI model for aerospace product integration," *Journal of Aviation and Aerospace Perspectives (JAAP)*, 2(2), 20–34.
Badiru, A. B. and M. Thomas (2013), Quantification of the PICK chart for process improvement decisions, *Journal of Enterprise Transformation*, 3(1), 1–15.

chapter three

Manufacturing cost analysis

Cost concepts and definitions

Cost management in a project environment refers to the functions required to maintain effective financial control of the project throughout its life cycle. There are several cost concepts that influence the economic aspects of managing projects. Within a given scope of analysis, there may be a combination of different types of cost aspects to consider. These cost aspects include the ones discussed below:

Actual Cost of Work Performed. This represents the cost actually incurred and recorded in accomplishing the work performed within a given time period.

Applied Direct Cost. This represents the amounts recognized in the time period associated with the consumption of labor, material, and other direct resources, without regard to the date of commitment or the date of payment. These amounts are to be charged to work-in-process (WIP) when resources are actually consumed, material resources are withdrawn from inventory for use, or material resources are received and scheduled for use within 60 days.

Budgeted Cost for Work Performed. This is the sum of the budgets for completed work plus the appropriate portion of the budgets for level of effort and apportioned effort. Apportioned effort is effort that by itself is not readily divisible into short-span work packages but is related in direct proportion to measured effort.

Budgeted Cost for Work Scheduled. This is the sum of budgets for all work packages and planning packages scheduled to be accomplished (including work-in-process), plus the amount of effort and apportioned effort scheduled to be accomplished within a given period of time.

Direct Cost. This is a cost that is directly associated with actual operations of a project. Typical sources of direct costs are direct material costs and direct labor costs. Direct costs are those that can be reasonably measured and allocated to a specific component of a project.

Economies of Scale. This refers to a reduction of the relative weight of the fixed cost in total cost by increasing output quantity. This helps to reduce the final unit cost of a product. Economies of scale is often simply referred to as the savings due to *mass production*.

Estimated Cost at Completion. This is the actual direct cost, plus indirect costs that can be allocated to the contract, plus the estimate of costs (direct and indirect) for authorized work remaining.

First Cost. This is the total initial investment required to initiate a project or the total initial cost of the equipment needed to start the project.

Fixed Cost. This is a cost incurred irrespective of the level of operation of a project. Fixed costs do not vary in proportion to the quantity of output. Example of costs that make up the fixed cost of a project are administrative expenses, certain types of taxes, insurance cost, depreciation cost, and debt servicing cost.

Incremental Cost. This refers to the additional cost of changing the production output from one level to another. Incremental costs are normally variable costs.

Indirect Cost. This is a cost that is indirectly associated with project operations. Indirect costs are those that are difficult to assign to specific components of a project. An example of an indirect cost is the cost of computer hardware and software needed to manage project operations. Indirect costs are usually calculated as a percentage of a component of direct costs. For example, the direct costs in an organization may be computed as 10% of direct labor costs.

Life-cycle Cost. This is the sum of all costs, recurring and nonrecurring, associated with a project during its entire life cycle.

Maintenance Cost. This is a cost that occurs intermittently or periodically for the purpose of keeping project equipment in good operating condition.

Marginal Cost. This is the additional cost of increasing production output by one additional unit. The marginal cost is equal to the slope of the total cost curve or line at the current operating level.

Operating Cost. This is a recurring cost needed to keep a project in operation during its life cycle. Operating costs may consist of such items as labor cost, material cost, and energy cost.

Opportunity Cost. This is the cost of forgoing the opportunity to invest in a venture that would have produced an economic advantage. Opportunity costs are usually incurred due to limited resources that make it impossible to take advantage of all investment opportunities. It is often defined as the cost of the best rejected opportunity. Opportunity costs can be incurred due to a missed opportunity rather than due to an intentional rejection. In many cases, opportunity costs are hidden or implied because they typically relate to future events that cannot be accurately predicted.

Overhead Cost. This is a cost incurred for activities performed in support of the operations of a project. The activities that generate overhead costs support the project efforts rather than contribute directly to the project goal. The handling of overhead costs varies widely from company to company. Typical overhead items are electric power cost, insurance premiums, cost of security, and inventory carrying cost.

Standard Cost. This is a cost that represents the normal or expected cost of a unit of the output of an operation. Standard costs are established in advance. They are developed as a composite of several component costs such as direct labor cost per unit, material cost per unit, and allowable overhead charge per unit.

Sunk Cost. This is a cost that occurred in the past and cannot be recovered under the present analysis. Sunk costs should have no bearing on the prevailing economic analysis and project decisions. Ignoring sunk costs is always a difficult task for analysts. For example, if $950,000 was spent four years ago to buy a piece of equipment for a technology-based project, a decision on whether or not to replace the equipment now should not consider that initial cost. But uncompromising analysts might find it difficult to ignore that much money. Similarly, an individual planning on selling a personal automobile would typically try to relate the asking price to what was paid for the automobile when it was acquired. This is wrong under the strict concept of sunk costs.

Total Cost. This is the sum of all the variable and fixed costs associated with a project.

Variable Cost. This is a cost that varies in direct proportion to the level of operation or quantity of output. For example, the costs of material and labor required to make an item will be classified as variable costs because they vary with changes in level of output.

Cost and cash flow analysis

The basic reason for performing economic analysis is to make a choice between mutually exclusive projects that are competing for limited resources. The cost performance of each project will depend on the timing and levels of its expenditures. The techniques of computing cash flow equivalence permit us to bring competing project cash flows to a common basis for comparison. The common basis depends on the prevailing interest rate. Two cash flows that are equivalent at a given interest rate will not be equivalent at a different interest rate. The basic techniques for converting cash flows from one point in time to another are presented in the next section.

Time value of money calculations

Cash flow conversion involves the transfer of project funds from one point in time to another. The following notation is used for the variables involved in the conversion process:

i = interest rate per period
n = number of interest periods
P = a present sum of money
F = a future sum of money
A = a uniform end-of-period cash receipt or disbursement
G = a uniform arithmetic gradient increase in period-by-period payments or disbursements.

In many cases, the interest rate used in performing economic analysis is set equal to the minimum attractive rate of return (MARR) of the decision maker. The MARR is also sometimes referred to as *hurdle rate, required internal rate of return* (IRR), *return on investment* (ROI), or *discount rate*. The value of MARR is chosen with the objective of maximizing the economic performance of a project.

Compound amount factor

The procedure for the single payment compound amount factor finds a future sum of money, F, that is equivalent to a present sum of money, P, at a specified interest rate, i, after n periods. This is calculated as

$$F = P(1+i)^n$$

Example   A sum of $5,000 is deposited in a project account and left there to earn interest for 15 years. If the interest rate per year is 12%, the compound amount after 15 years can be calculated as follows:

$$F = \$5000(1+0.12)^{15}$$

$$= \$27,367.85$$

Present worth factor

The present worth factor computes P when F is given. The present worth factor is obtained by solving for P in the equation for the compound amount factor. That is,

$$P = F(1+i)^{-n}$$

Suppose it is estimated that $15,000 would be needed to complete the implementation of a project five years from now. How much should be deposited in a special project fund now so that the fund would accrue to the required $15,000 exactly five years from now? If the special project fund pays interest at 9.2% per year, the required deposit would be

$$P = \$15000(1+0.092)^{-5}$$

$$= \$9660.03$$

Uniform series present worth factor

The uniform series present worth factor is used to calculate the present worth equivalent, P, of a series of equal end-of-period amounts, A. The derivation of the formula uses the finite sum of the present worth of the individual amounts in the uniform series cash flow as shown below.

$$P = \sum_{t=1}^{n} A(1+i)^{-1}$$

$$= A\left[\frac{(1+i)^n - 1}{i(1+i)^n}\right]$$

Example

Suppose the sum of $12,000 must be withdrawn from an account to meet the annual operating expenses of a multiyear project. The project account pays interest at 7.5% per year compounded on an annual basis. If the project is expected to last 10 years, how much must be deposited in the project account now so that the operating expenses of $12,000 can be withdrawn at the end of every year for 10 years? The project fund is expected to be depleted to zero by the end of the last year of the project. The first withdrawal will be made one year after the project account is opened, and no additional deposits will be made in the account during the project life cycle. The required deposit is calculated to be

$$P = \$12,000\left[\frac{(1+0.075)^{10} - 1}{0.075(1+0.075)^{10}}\right]$$

$$= \$82,368.92$$

Uniform series capital recovery factor

The capital recovery formula is used to calculate the uniform series of equal end-of-period payments, A, that are equivalent to a given present amount, P. This is the converse of the uniform series present amount factor. The equation for the uniform series capital recovery factor is obtained by solving for A in the uniform series present amount factor. That is,

$$A = P\left[\frac{i(1+i)^n}{(1+i)^n - 1}\right]$$

Example

Suppose a piece of equipment needed to launch a project must be purchased at a cost of $50,000. The entire cost is to be financed at 13.5% per year and repaid on a monthly installment schedule over four years. It is desired to calculate what the monthly loan payments will be. It is assumed that the first loan payment will be made exactly one month after the equipment is financed. If the interest rate of 13.5% per year is compounded monthly, then the interest rate per month will be 13.5%/12=1.125% per month. The number of interest periods over which the loan will be repaid is 4(12)=48 months. Consequently, the monthly loan payments are calculated to be

$$A = \$50,000\left[\frac{0.01125(1+0.01125)^{48}}{(1+0.01125)^{48} - 1}\right]$$

$$= \$1,353.82$$

Uniform series compound amount factor

The series compound amount factor is used to calculate a single future amount that is equivalent to a uniform series of equal end-of-period payments. Note that the future amount occurs at the same point in time as the last amount in the uniform series of payments. The factor is derived as shown below:

$$F = \sum_{t=1}^{n} A(1+i)^{n-1}$$

$$= A\left[\frac{(1+i)^n - 1}{i}\right]$$

Example

If equal end-of-year deposits of $5,000 are made to a project fund paying 8% per year for 10 years, how much can be expected to be available for withdrawal from the account for capital expenditure immediately after the last deposit is made?

$$F = \$5000\left[\frac{(1+0.08)^{10} - 1}{0.08}\right]$$

$$= \$72,432.50$$

Uniform series sinking fund factor

The sinking fund factor is used to calculate the uniform series of equal end-of-period amounts, A, that are equivalent to a single future amount, F. This is the reverse of the uniform series compound amount factor. The formula for the sinking fund is obtained by solving for A in the formula for the uniform series compound amount factor. That is,

$$A = F\left[\frac{i}{(1+i)^n - 1}\right]$$

Example

How large are the end-of-year equal amounts that must be deposited into a project account so that a balance of $75,000 will be available for withdrawal immediately after the twelfth annual deposit is made? The initial balance in the account is zero at the beginning of the first year. The account pays 10% interest per year. Using the formula for the sinking fund factor, the required annual deposits are

$$A = \$75,000\left[\frac{0.10}{(1+0.10)^{12} - 1}\right]$$

$$= \$3,507.25$$

Capitalized cost formula

Capitalized cost refers to the present value of a single amount that is equivalent to a perpetual series of equal end-of-period payments. This is an extension of the series present worth factor with an infinitely large number of periods.

Using the limit theorem from calculus as n approaches infinity, the series present worth factor reduces to the following formula for the capitalized cost:

$$P = \lim_{n \to \infty} A\left[\frac{(1+i)^n - 1}{i(1+i)^n}\right]$$

$$= A\left\{\lim_{n \to \infty}\left[\frac{(1+i)^n - 1}{i(1+i)^n}\right]\right\}$$

$$= A\left(\frac{1}{i}\right)$$

Example

How much should be deposited in a general fund to service a recurring public service project to the tune of $6,500 per year forever if the fund yields an annual interest rate of 11%? Using the capitalized cost formula, the required one-time deposit to the general fund is

$$P = \frac{\$6500}{0.11}$$

$$= \$59,090.91$$

The formulas presented above represent the basic cash flow conversion factors. The factors are widely tabulated, for convenience, in engineering economy books. Several variations and extensions of the factors are available. Such extensions include the arithmetic gradient series factor and the geometric series factor. Variations in the cash flow profiles include situations where payments are made at the beginning of each period rather than at the end and situations where a series of payments contains unequal amounts. Conversion formulas can be derived mathematically for those special cases by using the basic factors presented above.

Arithmetic gradient series

The gradient series cash flow involves an increase of a fixed amount in the cash flow at the end of each period. Thus, the amount at a given point in time is greater than the amount at the preceding period by a constant amount. This constant amount is denoted by G. The size of the cash flow in the gradient series at the end of period t is calculated as

$$A_t = (t-1)G, \quad t = 1, 2, \ldots, n$$

The total present value of the gradient series is calculated by using the present amount factor to convert each individual amount from time t to time 0 at an interest rate of $i\%$ per period and summing up the resulting present values. The finite summation reduces to a closed form as shown below:

$$P = \sum_{t=1}^{n} A_t (1+i)^{-t}$$

$$= \sum_{t=1}^{n} (t-1)G(1+i)^{-t}$$

$$= G \sum_{t=1}^{n} (t-1)(1+i)^{-t}$$

$$= G \left[\frac{(1+i)^n - (1+ni)}{i^2(1+i)^n} \right]$$

Example

The cost of supplies for a 10-year period increases by $1,500 every year starting at the end of year two. There is no supplies cost at the end of the first year. If interest rate is 8% per year, determine the present amount that must be set aside at time zero to take care of all the future supplies expenditures. We have $G=1500$, $i=0.08$, and $n=10$. Using the arithmetic gradient formula, we obtain

$$P = 1500 \left[\frac{1-(1+10(0.08))(1+0.08)^{-10}}{(0.08)^2} \right]$$

$$= \$1500(25.9768)$$

$$= \$38,965.20$$

In many cases, an arithmetic gradient starts with some base amount at the end of the first period and then increases by a constant amount thereafter. The non-zero base amount is denoted as A_1.

The calculation of the present amount for such cash flows requires breaking the cash flow into a uniform series cash flow of amount A_1 and an arithmetic gradient cash flow with zero base amount. The uniform series present worth formula is used to calculate the present worth of the uniform series portion, while the basic gradient series formula is used to calculate the gradient portion. The overall present worth is then calculated as

$$P = P_{\text{uniform series}} + P_{\text{gradient series}}$$

$$= A_1 \left[\frac{(1+i)^n - 1}{i(1+i)^n} \right] + G \left[\frac{(1+i)^n - (1+ni)}{i^2(1+i)^n} \right]$$

Increasing geometric series cash flow

In an increasing geometric series cash flow, the amounts in the cash flow increase by a constant percentage from period to period. There is a positive base amount, A_1, at the end of period 1. The amount at time t is denoted as

$$A_t = A_{t-1}(1+j), \quad t = 2,3,\dots,n$$

where j is the percentage increase in the cash flow from period to period. By doing a series of back substitutions, we can represent A_t in terms of A_1 instead of in terms of A_{t-1} as shown below:

$$A_2 = A_1(1+j)$$

$$A_3 = A_2(1+j) = A_1(1+j)(1+j)$$

$$\dots$$

$$A_t = A_1(1+j)^{t-1}, \quad t = 1,2,3,\dots,n$$

The formula for calculating the present worth of the increasing geometric series cash flow is derived by summing the present values of the individual cash flow amounts. That is,

$$P = \sum_{t=1}^{n} A_t(1+i)^{-t}$$

$$= \sum_{t=1}^{n} \left[A_1(1+j)^{t-1} \right](1+i)^{-t}$$

$$= \frac{A_1}{(1+j)} \sum_{t=1}^{n} \left(\frac{1+j}{1+i} \right)^t$$

$$= A_1 \left[\frac{1-(1+j)^n(1+i)^{-n}}{i-j} \right], \qquad i \neq j$$

If $i=j$, the formula above reduces to the limit as $i \rightarrow j$, shown below:

$$P = \frac{nA_1}{1+i}, \qquad i=j$$

Example

Suppose funding for a five-year project is to increase by 6% every year with an initial funding of $20,000 at the end of the first year. Determine how much must be deposited into a budget account at time zero in order to cover the anticipated funding levels if the budget account pays 10% interest per year. We have $j=6\%$, $i=10\%$, $n=5$, $A_1=\$20,000$. Therefore,

$$P = 20,000 \left[\frac{1-(1+0.06)^5(1+0.10)^{-5}}{0.10-0.06} \right]$$

$$= \$20,000(4.2267)$$

$$= \$84,533.60$$

Decreasing geometric series cash flow

In a decreasing geometric series cash flow, the amounts in the cash flow decrease by a constant percentage from period to period. The cash flow starts at some positive base amount, A_1, at the end of period 1. The amount of time t is denoted as

$$A_t = A_{t-1}(1-j), \qquad t = 2,3,\ldots,n$$

where j is the percentage decrease in the cash flow from period to period. As in the case of the increasing geometric series, we can represent A_t in terms of A_1:

$$A_2 = A_1(1-j)$$

$$A_3 = A_2(1-j) = A_1(1-j)(1-j)$$

$$\cdots$$

$$A_t = A_1(1-j)^{t-1}, \quad t = 1,2,3,\ldots,n$$

The formula for calculating the present worth of the decreasing geometric series cash flow is derived by finite summation as in the case of the increasing geometric series. The final formula is

$$P = A_1 \left[\frac{1-(1-j)^n(1+i)^{-n}}{i+j} \right]$$

Example

The contract amount for a three-year project is expected to decrease by 10% every year with an initial contract of $100,000 at the end of the first year. Determine how much must be available in a contract reservoir fund at time zero in order to cover the contract amounts. The fund pays 10% interest per year. Because $j=10\%$, $i=10\%$, $n=3$, and $A_1=\$100,000$, we should have

$$P = 100,000 \left[\frac{1-(1-0.10)^3(1+0.10)^{-3}}{0.10+0.10} \right]$$

$$= \$100,000(2.2615)$$

$$= \$226,150$$

Internal rate of return

The internal rate of return (IRR) for a cash flow is defined as the interest rate that equates the future worth at time n or present worth at time 0 of the cash flow to zero. If we let i^* denote the internal rate of return, then we have

$$FW_{t=n} = \sum_{t=0}^{n} (\pm A_t)(1+i^*)^{n-t} = 0$$

$$PW_{t=0} = \sum_{t=0}^{n} (\pm A_t)(1+i^*)^{-t} = 0$$

where "+" is used in the summation for positive cash flow amounts or receipts and "—" is used for negative cash flow amounts or disbursements. A_t denotes the cash flow amount at time t, which may be a receipt (+) or a disbursement (—). The value of i^* is referred to as *discounted cash flow rate of return, internal rate of return,* or *true rate of return.* The procedure above essentially calculates the net future worth or the net present worth of the cash flow. That is,

$$\text{Net future worth} = \text{Future worth of receipts}$$

$$-\text{Future worth of disbursements}$$

$$\text{NFW} = \text{FW}_{\text{receipts}} - \text{FW}_{\text{disbursements}}$$

$$\text{Net present worth} = \text{Present worth of receipts}$$

$$-\text{Present worth of disbursements}$$

$$\text{NPW} = \text{PW}_{\text{receipts}} - \text{PW}_{\text{disbursements}}$$

Setting the NPW or NFW equal to zero and solving for the unknown variable i determines the internal rate of return of the cash flow.

Benefit-cost ratio

The benefit-cost ratio of a cash flow is the ratio of the present worth of benefits to the present worth of costs. This is defined as

$$B/C = \frac{\displaystyle\sum_{t=0}^{n} B_t(1+i)^{-t}}{\displaystyle\sum_{t=0}^{n} C_t(1+i)^{-t}}$$

$$= \frac{\text{PW}_{\text{benefits}}}{\text{PW}_{\text{costs}}}$$

where B_t is the benefit (receipt) at time t and C_t is the cost (disbursement) at time t. If the benefit-cost ratio is greater than 1, then the investment is acceptable. If the ratio is less than 1, the investment is not acceptable. A ratio of 1 indicates a break-even situation for the project.

Simple payback period

Payback period refers to the length of time it will take to recover an initial investment. The approach does not consider the impact of the time value of money. Consequently, it is not an accurate method of evaluating the worth of an investment. However, it is a simple technique that is used widely to perform a "quick-and-dirty" assessment of investment performance. Also, the technique considers only the initial cost. Other costs that may occur after time zero are not included in the calculation. The payback period is defined as the smallest value of n (n_{min}) that satisfies the following expression:

$$\sum_{t=1}^{n_{min}} R_t \geq C_0$$

where R_t is the revenue at time t and C_0 is the initial investment. The procedure calls for a simple addition of the revenues period by period until enough total has been accumulated to offset the initial investment.

Example
An organization is considering installing a new computer system that will generate significant savings in material and labor requirements for order processing. The system has an initial cost of $50,000. It is expected to save the organization $20,000 a year. The system has an anticipated useful life of five years with a salvage value of $5,000. Determine how long it would take for the system to pay for itself from the savings it is expected to generate. Because the annual savings are uniform, we can calculate the payback period by simply dividing the initial cost by the annual savings. That is,

$$n_{min} = \frac{\$50,000}{\$20,000}$$

$$= 2.5\,years$$

Note that the salvage value of $5,000 is not included in the above calculation because the amount is not realized until the end of the useful life of the asset (i.e., after five years). In some cases, it may be desired to consider the salvage value. In that case, the amount to be offset by the annual savings will be the net cost of the asset. In that case, we would have

$$n_{min} = \frac{\$50,000 - \$5000}{\$20,000}$$

$$= 2.25\,years$$

If there are tax liabilities associated with the annual savings, those liabilities must be deducted from the savings before calculating the payback period.

Discounted payback period
In this book, we introduce the *discounted payback period* approach in which the revenues are reinvested at a certain interest rate. The payback period

is determined when enough money has been accumulated at the given interest rate to offset the initial cost as well as other interim costs. In this case, the calculation is done by the following expression:

$$\sum_{t=1}^{n_{min}} R_t(1+i)^{n_{min}-1} \geq \sum_{t=0}^{n_{min}} C_t$$

Example

A new solar cell unit is to be installed in an office complex at an initial cost of $150,000. It is expected that the system will generate annual cost savings of $22,500 on the electricity bill. The solar cell unit will need to be overhauled every five years at a cost of $5,000 per overhaul. If the annual interest rate is 10%, find the *discounted payback period* for the solar cell unit considering the time value of money. The costs of overhaul are to be considered in calculating the discounted payback period.

Using the single payment compound amount factor for one period iteratively, the following solution is obtained:

Time period 1: $22,500
Time period 2: $22,500 + $22,500(1.10)1 = $47,250
Time period 3: $22,500 + $47,250(1.10)1 = $74,475
Time period 4: $22,500 + $74,475(1.10)1 = $104,422.50
Time period 5: $22,500 + $104,422.50(1.10)1 − $5000 = $132,364.75
Time period 6: $22,500 + $132,364.75(1.10)1 = $168,101.23

The initial investment is $150,000. By the end of period 6, we have accumulated $168,101.23, more than the initial cost. Interpolating between period 5 and period 6, we obtain

$$n_{min} = 5 + \frac{150,000 - 132,364.75}{168,101.23 - 132,364.75}(6-5)$$

$$= 5.49$$

That is, it will take 5.49 years, or five years and six months, to recover the initial investment.

Investment life for multiple returns

The time it takes an amount to reach a certain multiple of its initial level is often of interest in many investment scenarios. The "Rule of 72" is one simple approach to calculating how long it will take an investment to double in value at a given interest rate per period. The Rule of 72 gives the following formula for estimating the doubling period:

$$n = \frac{72}{i}$$

where i is the interest rate expressed in percentage. Referring to the single payment compound amount factor, we can set the future amount equal to twice the present amount and then solve for n, the number of periods. That is, $F = 2P$. Thus,

$$2P = P(1+i)^n$$

Solving for n in the above equation yields an expression for calculating the exact number of periods required to double P:

$$n = \frac{1n(2)}{1n(1+i)}$$

where i is the interest rate expressed in decimals. In the general case, for exact computation, the length of time it would take to accumulate m multiple of P is expressed as

$$n = \frac{1n(m)}{1n(1+i)}$$

where m is the desired multiple. For example, at an interest rate of 5% per year, the time it would take an amount, P, to double in value ($m = 2$) is 14.21 years. This, of course, assumes that the interest rate will remain constant throughout the planning horizon.

Effects of inflation

Inflation is a major player in financial and economic analyses of projects. Multiyear projects are particularly subject to the effects of inflation. Inflation can be defined as the decline in purchasing power of money.

Some of the most common causes of inflation are:

- Increase in amount of currency in circulation
- Shortage of consumer goods
- Escalation of the cost of production
- Arbitrary increase of prices by resellers

The general effects of inflation are felt in terms of increase in the prices of goods and decrease in the worth of currency. In cash flow analysis, return on investment (ROI) for a project will be affected by time value of money as well as inflation. The *real interest rate* (*d*) is defined as the desired rate of return in the absence of inflation. When we talk of "today's dollars" or "constant dollars," we are referring to the use of real interest rate. *Combined interest rate* (*i*) is the rate of return combining real interest rate and inflation rate. If we denote the *inflation rate* as *j*, then the relationship between the different rates can be expressed as

$$1 + i = (1 + d)(1 + j)$$

Thus, the combined interest rate can be expressed as

$$i = d + j + dj$$

Note that if $j = 0$ (i.e., no inflation), then $i = d$. We can also define *commodity escalation rate* (*g*) as the rate at which individual commodity prices escalate. This may be greater than or less than the overall inflation rate. In practice, several measures are used to convey inflationary effects. Some of these are *consumer price index*, *producer price index*, and *wholesale price index*. A *"market basket"* rate is defined as the estimate of inflation based on a weighted average of the annual rates of change in the costs of a wide range of representative commodities. A "then-current" cash flow is a cash flow that explicitly incorporates the impact of inflation. A "constant worth" cash flow is a cash flow that does not incorporate the effect of inflation. The real interest rate, *d*, is used for analyzing constant worth cash flows.

The then-current cash flow is the equivalent cash flow considering the effect of inflation. C_k is what it would take to buy a certain "basket" of goods after *k* time periods if there was no inflation. T_k is what it would take to buy the same "basket" in *k* time period if inflation was taken into account. For the constant worth cash flow, we have

$$C_k = T_0, \quad k = 1, 2, \ldots, n$$

and for the then-current cash flow, we have

$$T_k = T_0 (1 + j)^k, \quad k = 1, 2, \ldots, n$$

where *j* is the inflation rate. If $C_k = T_0 = \$100$ under the constant worth cash flow, then we mean \$100 worth of buying power. If we are using the commodity escalation rate, *g*, then we will have

$$T_k = T_0 (1 + g)^k, \quad k = 1, 2, \ldots, n$$

Thus, a then-current cash flow may increase based on both a regular inflation rate (j) and a commodity escalation rate (g). We can convert a then-current cash flow to a constant worth cash flow by using the following relationship:

$$C_k = T_k (1+j)^{-k}, \quad k = 1, 2, \ldots, n$$

If we substitute T_k from the commodity escalation cash flow into the expression for C_k above, we get

$$C_k = T_k(1+j)^{-k}$$
$$= T_0(1+g)^k(1+j)^{-k}$$
$$= T_0[(1+g)/(1+j)]^k, \quad k = 1, 2, \ldots, n$$

Note that if $g=0$ and $j=0$, the $C_k = T_0$. That is, no inflationary effect. We now define effective commodity escalation rate (v) as

$$v = \left[(1+g)/(1+j)\right] - 1$$

and we can express the commodity escalation rate (g) as

$$g = v + j + vj.$$

Inflation can have a significant impact on the financial and economic aspects of a project. Inflation may be defined, in economic terms, as the increase in the amount of currency in circulation, resulting in a relatively high and sudden fall in its value. To a producer, inflation means a sudden increase in the cost of items that serve as inputs for the production process (equipment, labor, materials, etc.). To the retailer, inflation implies an imposed higher cost of finished products. To an ordinary citizen, inflation portends an unbearable escalation of prices of consumer goods. All these views are interrelated in a project management environment.

The amount of money supply, as a measure of a country's wealth, is controlled by the government. With no other choice, governments often feel impelled to create more money or credit to take care of old debts and pay for social programs. When money is generated at a faster rate than the growth of goods and services, it becomes a surplus commodity, and its value (purchasing power) will fall. This means that there will be too much money available to buy only a few goods and services. When the purchasing power of a currency falls, each individual in a product's life cycle has to dispense more of the currency in order to obtain the product. Some of the classic concepts of inflation are discussed below:

1. Increases in producer's costs are passed on to consumers. At each stage of the product's journey from producer to consumer, prices are escalated disproportionately in order to make a good profit. The overall increase in the product's price is directly proportional to the number of intermediaries it encounters on its way to the consumer. This type of inflation is called *cost-driven (or cost-push) inflation.*
2. Excessive spending power of consumers forces an upward trend in prices. This high spending power is usually achieved at the expense of savings. The law of supply and demand dictates that the more the demand, the higher the price. This type of inflation is known as *demand-driven (or demand-pull) inflation.*
3. Impact of international economic forces can induce inflation in a local economy. Trade imbalances and fluctuations in currency values are notable examples of international inflationary factors.
4. Increasing base wages of workers generates more disposable income and, hence, higher demands for goods and services. The high demand, consequently, creates a rise on prices. Coupled with this, employers pass on the additional wage cost to consumers through higher prices. This type of inflation is perhaps the most difficult to solve because wages set by union contracts and prices set by producers almost never fall – at least not permanently. This type of inflation may be referred to as *wage-driven (or wage-push) inflation.*
5. Easy availability of credit leads consumers to "buy now and pay later" and, thereby, creates another loophole for inflation. This is a dangerous type of inflation because the credit not only pushes prices up, but it also leaves consumers with less money later on to pay for the credit. Eventually, many credits become uncollectible debts, which may then drive the economy into recession.
6. Deficit spending results in an increase in money supply and, thereby, creates less room for each dollar to get around. The popular saying, "a dollar does not go far anymore," simply refers to inflation in layman's terms. The different levels of inflation may be categorized as discussed below.

Mild inflation

When inflation is mild (2%–4%), the economy actually prospers. Producers strive to produce at full capacity in order to take advantage of the high prices to the consumer. Private investments tend to be brisk and more jobs become available. However, the good fortune may only be temporary. Prompted by the prevailing success, employers are tempted to seek larger profits and workers begin to ask for higher wages. They cite their employer's prosperous business as a reason to bargain for bigger shares of the business profit. Thus, we end up with a vicious cycle where the producer asks for higher prices, the unions ask for higher wages, and inflation starts an upward trend.

Moderate inflation

Moderate inflation occurs when prices increase at 5%–9%. Consumers start purchasing more as an edge against inflation. They would rather spend their money now than watch it decline further in purchasing power. The increased market activity serves to fuel further inflation.

Severe inflation

Severe inflation is indicated by price escalations of 10% or more. Double-digit inflation implies that prices rise much faster than wages do. Debtors tend to be the ones who benefit from this level of inflation because they repay debts with money that is less valuable then the money borrowed.

Hyperinflation

When each price increase signals the increase in wages and costs, which again sends prices further up, the economy has reached a stage of malignant galloping inflation or hyperinflation. Rapid and uncontrollable inflation destroys the economy. The currency becomes economically useless as the government prints it excessively to pay for obligations.

Inflation can affect any project in terms of raw materials procurement, salaries and wages, and/or cost tracking dilemma. Some effects are immediate and easily observable. Other effects are subtle and pervasive. Whatever form it takes, inflation must be considered in long-term project planning and control. Large projects may be adversely affected by the effects of inflation in terms of cost overruns and poor resource utilization. The level of inflation will determine the severity of the impact on projects.

Break-even analysis

Break-even analysis refers to the determination of the balanced performance level where project income is equal to project expenditure. The total cost of an operation is expressed as the sum of the fixed and variable costs with respect to output quantity. That is,

$$TC(x) = FC + VC(x)$$

where x is the number of units produced, $TC(x)$ is the total cost of producing x units, FC is the total fixed cost, and $VC(x)$ is the total variable cost associated with producing x units. The total revenue resulting from the sale of x units is defined as

$$TR(x) = px$$

where p is the price per unit. The profit due to the production and sale of x units of the product is calculated as

$$P(x) = TR(x) - TC(x)$$

The break-even point of an operation is defined as the value of a given parameter that will result in neither profit nor loss. The parameter of interest may be the number of units produced, the number of hours of operation, the number of units of a resource type allocated, or any other measure of interest. At the break-even point, we have the following relationship:

$$TR(x) = TC(x) \text{ or } P(x) = 0$$

In some cases, there may be a known mathematical relationship between cost and the parameter of interest. For example, there may be a linear cost relationship between the total cost of a project and the number of units produced. The cost expressions facilitate straightforward break-even analysis. When two project alternatives are compared, the break-even point refers to the point of indifference between the two alternatives. The variable $x1$ represents the point where projects A and B are equally desirable, $x2$ represents where A and C are equally desirable, and $x3$ represents where B and C are equally desirable. The analysis shows that if we are operating below a production level of $x2$ units, then project C is the preferred project among the three. If we are operating at a level more than $x2$ units, then project A is the best choice.

Example

Three project alternatives are being considered for producing a new product. The required analysis involves determining which alternative should be selected on the basis of how many units of the product are produced per year. Based on past records, there is a known relationship between the number of units produced per year, x, and the net annual profit, $P(x)$, from each alternative. The level of production is expected to be between 0 and 250 units per year. The net annual profits (in thousands of dollars) are given below for each alternative:

Project A: $P(x) = 3x - 200$
Project B: $P(x) = x$
Project C: $P(x) = (1/50)x^2 - 300$

This problem can be solved mathematically by finding the intersection points of the profit functions and evaluating the respective profits over the given range of product units. It can also be solved by a graphical approach. Such a plot is called a *break-even chart*. A review of the calculations shows that Project B should be selected if between 0 and 100 units are to be produced. Project A should be selected if between 100 and 178.1 units (178 physical units) are to be produced. Project C should be selected if more than 178 units are to be produced. It should be noted that if less than 66.7 units (66 physical units) are produced, Project A will generate net loss rather than net profit. Similarly, Project C will generate losses if less than 122.5 units (122 physical units) are produced.

Profit ratio analysis

Break-even charts offer opportunities for several different types of analysis. In addition to the break-even points, other measures of worth or criterion may be derived from the charts. A measure, called *profit ratio* (Badiru and Omitaomu, 2007), is presented here for the purpose of obtaining a further comparative basis for competing projects. Profit ratio is defined as the ratio of the profit area to the sum of the profit and loss areas in a break-even chart. That is,

$$\text{Profit ratio} = \frac{\text{Area of profit region}}{\text{Area of profit region} + \text{Area of loss region}}$$

For example, suppose the expected revenue and the expected total cost associated with a project are given, respectively, by the following expressions:

$$R(x) = 100 + 10x$$

$$TC(x) = 2.5x + 250$$

where x is the number of units produced and sold from the project. The break-even point is shown to be 20 units. Net profits are realized from the project if more than 20 units are produced, and net losses are realized if less than 20 units are produced. It should be noted that the revenue function represents an unusual case where a revenue of $100 is realized when zero units are produced.

Suppose it is desired to calculate the profit ratio for this project if the number of units that can be produced is limited to between 0 and

100 units. The surface area of the profit region and the area of the loss region can be calculated by using the standard formula for finding the area of a triangle: Area = (1/2)(Base)(Height). Using this formula, we have the following:

$$\text{Area of profit region} = \frac{1}{2}(\text{Base})(\text{Height})$$

$$= \frac{1}{2}(1100-500)(100-20)$$

$$= 24,000 \text{ square units}$$

$$\text{Area of loss region} = \frac{1}{2}(\text{Base})(\text{Height})$$

$$= \frac{1}{2}(250-100)(20)$$

$$= 1500 \text{ square units}$$

Thus, the profit ratio is computed as

$$\text{Profit ratio} = \frac{24,000}{24,000+1500}$$

$$= 0.9411$$

$$= 94.11\%$$

The profit ratio may be used as a criterion for selecting among project alternatives. If this is done, the profit ratios for all the alternatives must be calculated over the same values of the independent variable. The project with the highest profit ratio will be selected as the desired project. Both the revenue and cost functions for the project are non-linear. The revenue and cost are defined as follows:

$$R(x) = 160x - x^2$$

$$TC(x) = 500 + x^2$$

If the cost and/or revenue functions for a project are not linear, the areas bounded by the functions may not be easily determined. For those cases, it may be necessary to use techniques such as definite integrals to find the areas. The computations indicate that the project generates a loss if less than 3.3 units (3 actual units) or more than 76.8 (76 actual units) are produced. The respective profit and loss areas on the chart are calculated as follows:

$$\text{Area 1 (loss)} = \int_0^{3.3} \left[\left(500 + x^2 \right) - \left(160x - x^2 \right) \right] dx$$

$$= 802.8 \text{ unit-dollars}$$

$$\text{Area 2 (profit)} = \int_{3.3}^{76.8} \left[\left(160x - x^2 \right) - \left(500 + x^2 \right) \right] dx$$

$$= 132,272.08 \text{ unit-dollars}$$

$$\text{Area 3 (loss)} = \int_{76.8}^{100} \left[\left(500 + x^2 \right) - \left(160x - x^2 \right) \right] dx$$

$$= 48,135.98 \text{ unit-dollars}$$

Consequently, the profit ratio for Project B is computed as

$$\text{Profit ratio} = \frac{\text{Total area of profit region}}{\text{Total area of profit region} + \text{Total area of loss region}}$$

$$\frac{132,272.08}{802.76 + 132,272.08 + 48,135.98}$$

$$= 0.7299$$

$$= 72.99\%$$

The profit ratio approach evaluates the performance of each alternative over a specified range of operating levels. Most of the existing evaluation methods use single-point analysis with the assumption that the operating condition is fixed at a given production level. The profit ratio measure allows an analyst to evaluate the net yield of an alternative given that the production level may shift from one level to another. An alternative, for example, may operate at a loss for most of its early life, while it may generate large incomes to offset the losses in its later stages. Conventional methods cannot easily capture this type of transition from one performance level to another. In addition to being used to compare alternate projects, the profit ratio may also be used for evaluating the economic feasibility of a single project. In such a case, a decision rule may be developed. An example of such a decision rule is:

If profit ratio is greater than 75%, accept the project.
If profit ratio is less than or equal to 75%, reject the project.

Amortization analysis

Many capital investment projects are financed with external funds. A careful analysis must be conducted to ensure that the amortization schedule can be handled by the organization involved. A computer program such as GAMPS (graphic evaluation of amortization payments) might be used for this purpose (Badiru, 2016). The program analyzes the installment payments, the unpaid balance, principal amounts paid per period, total installment payment, and current cumulative equity. It also calculates the "equity break-even point" (Badiru, 2016) for the debt being analyzed. The equity break-even point indicates the time when the unpaid balance on a loan is equal to the cumulative equity on the loan. With the output of this program, the basic cost of servicing the project debt can be evaluated quickly. A part of the output of the program presents the percentage of the installment payment going into equity and interest charge respectively. The computational procedure for analyzing project debt follows the steps below:

1. Given a principal amount, P, a periodic interest rate, i (in decimals), and a discrete time span of n periods, the uniform series of equal end-of-period payments needed to amortize P is computed as

$$A = \frac{P[i(1+i)^n]}{(1+i)^n - 1}$$

   It is assumed that the loan is to be repaid in equal monthly payments. Thus, $A(t) = A$, for each period t throughout the life of the loan.

2. The unpaid balance after making t installment payments is given by

$$U(t) = \frac{A[1-(1+i)^{t-n}]}{i}$$

3. The amount of equity or principal amount paid with installment payment number t is given by

$$E(t) = A(1+i)^{t-n-1}$$

4. The amount of interest charge contained in installment payment number t is derived to be

$$I(t) = A[1-(1+i)^{t-n-1}]$$

where $A = E(t) + I(t)$.

5. The cumulative total payment made after t periods is denoted by

$$C(t) = \sum_{k=1}^{t} A(k)$$

$$= \sum_{k=1}^{t} A$$

$$= (A)(t)$$

6. The cumulative interest payment after t periods is given by

$$Q(t) = \sum_{x=1}^{t} I(x)$$

7. The cumulative principal payment after t periods is computed as

$$S(t) = \sum_{k=1}^{t} E(k)$$

$$= A \sum_{k=1}^{t} (1+i)^{-(n-k+1)}$$

$$= A \left[\frac{(1+i)^t - 1}{i(1+i)^n} \right]$$

where:

$$\sum_{n-1}^{t} x^n = \frac{x^{t+1} - x}{x - 1}$$

8. The percentage of interest charge contained in installment payment number t is

$$f(t) = \frac{I(t)}{A} (100\%)$$

9. The percentage of cumulative interest charge contained in the cumulative total payment up to and including payment number t is

$$F(t) = \frac{Q(t)}{C(t)} (100\%)$$

10. The percentage of cumulative principal payment contained in the cumulative total payment up to and including payment number t is

$$H(t) = \frac{S(t)}{C(t)}$$

$$= \frac{C(t) - Q(t)}{C(t)}$$

$$= 1 - \frac{Q(t)}{C(t)}$$

$$= 1 - F(t)$$

Example

Suppose that a manufacturing productivity improvement project is to be financed by borrowing $500,000 from an industrial development bank. The annual nominal interest rate for the loan is 10%. The loan is to be repaid in equal monthly installments over a period of 15 years. The first payment on the loan is to be made exactly one month after financing is approved. It is desired to perform a detailed analysis of the loan schedule.

The tabulated result shows a monthly payment of $5,373.04 on the loan. Considering time $t = 10$ months, one can see the following results:

$U(10) = \$487,475.13$ (unpaid balance)

$A(10) = \$5373.04$ (monthly payment)

$E(10) = \$1299.91$ (equity portion of the tenth payment)

$I(10) = \$4073.13$ (interest charge contained in the tenth payment)

$C(10) = \$53,730.40$ (total payment to date)

$S(10) = \$12,526.21$ (total equity to date)

$f(10) = 75.81\%$ (percentage of the tenth payment going into interest charge)

$F(10) = 76.69\%$ (percentage of the total payment going into interest charge)

Thus, over 76% of the sum of the first 10 installment payments goes into interest charges. The analysis shows that by time $t = 180$, the unpaid balance has been reduced to zero. That is, $U(180) = 0.0$. The total payment made on the loan is $967,148.40 and the total interest charge is $967,148.20 − $500,000 = $467,148.20. Thus, 48.30% of the total payment goes into interest charges. The information about interest charges might be very useful for tax purposes. The tabulated output, if used, would show that equity builds up slowly while unpaid balance decreases slowly. Note that very little equity is accumulated during the first three years of the loan schedule. The effects of inflation, depreciation, property appreciation, and other economic factors are not included in the analysis presented above. A project analyst should include such factors whenever they are relevant to the loan situation.

The point at which the curves intersect is referred to as the *equity break-even point*. It indicates when the unpaid balance is exactly equal to the accumulated equity or the cumulative principal payment. For the example, the equity break-even point is 120.9 months (over 10 years). The importance of the equity break-even point is that any equity accumulated after that point represents the amount of ownership or equity that the debtor is entitled to after the unpaid balance on the loan is settled with project collateral. The implication of this is very important, particularly in the case of mortgage loans. "Mortgage" is a word with French origin, meaning *death pledge* – perhaps a sarcastic reference to the burden of mortgage loans. The equity break-even point can be calculated directly from the formula derived below:

Let the equity break-even point, x, be defined as the point where $U(x) = S(x)$. That is,

$$A\left[\frac{1-(1+i)^{-(n-x)}}{i}\right] = A\left[\frac{(1+i)^{x}-1}{i(1+i)^{n}}\right]$$

Multiplying both the numerator and denominator of the left-hand side of the above expression by $(1+i)^{n}$ and simplifying yields

$$\frac{(1+i)^{n}-(1+i)^{x}}{i(1+i)^{n}}$$

on the left-hand side. Consequently, we have

$$(1+i)^{n}-(1+i)^{x} = (1+i)^{x}-1$$

$$(1+i)^{x} = \frac{(1+i)^{n}+1}{2}$$

which yields the equity break-even expression:

$$x = \frac{\ln[0.5(1+i)^{n}+0.5]}{\ln(1+i)}$$

where:

    ln   is the natural log function
    n is the number of periods in the life of the loan
    i is the interest rate per period.

The total payment starts from $0.0 at time zero and goes up to $967,147.20 by the end of the last month of the installment payments. Because only $500,000 was borrowed, the total interest payment on the loan is $967,147.20 − $500,00 = $467,147.20. The cumulative principal payment starts at $0.0 at time zero and slowly builds up to $500,001.34, which is the original loan amount. The extra $1.34 is due to round-off error in the calculations.

The percentage of interest charge in the monthly payments starts at 77.55% for the first month and decreases to 0.83% for the last month. By comparison, the percentage of interest in the total payment starts also at 77.55% for the first month and slowly decreases to 48.30% by the time the last payment is made at time 180. It is noted that an increasing proportion of the monthly payment goes into the principal payment as time goes on. If the interest charges are tax deductible, the decreasing values of $f(t)$ mean that there would be decreasing tax benefits from the interest charges in the later months of the loan.

Manufacturing cost estimation

Cost estimation and budgeting help establish a strategy for allocating resources in project planning and control. There are three major categories of cost estimation for budgeting. These are based on the desired level of accuracy. The categories are *order-of-magnitude estimates, preliminary cost estimates,* and *detailed cost estimates.* Order-of-magnitude cost estimates are usually gross estimates based on the experience and judgment of the estimator. They are sometimes called "ballpark" figures. These estimates are typically made without a formal evaluation of the details involved in the project. The level of accuracy associated with order-of-magnitude estimates can range from −50% to +50% of the actual cost. These estimates provide a quick way of getting cost information during the initial stages of a project.

$$50\%\left(\text{Actual cost}\right) \leq \text{Order-of-magnitude estimate} \leq 150\%\left(\text{Actual cost}\right)$$

Preliminary cost estimates are also gross estimates, but with a higher level of accuracy. In developing preliminary cost estimates, more attention is paid to some selected details of the project. An example of a preliminary cost estimate is the estimation of expected labor cost. Preliminary estimates are useful for evaluating project alternatives before final commitments are made. The level of accuracy associated with preliminary estimates can range from −20% to +20% of the actual cost.

$$80\%\left(\text{Actual cost}\right) \leq \text{Preliminary estimate} \leq 120\%\left(\text{Actual cost}\right)$$

Detailed cost estimates are developed after careful consideration is given to all the major details of a project. Because of the considerable amount of time and effort needed to develop detailed cost estimates, the estimates are usually developed after there is firm commitment that the project will take off. Detailed cost estimates are important for evaluating actual cost performance during the project. The level of accuracy associated with detailed estimates normally range from −5% to +5% of the actual cost.

$$95\%\left(\text{Actual cost}\right) \leq \text{Detailed cost} \leq 105\%\left(\text{Actual cost}\right)$$

There are two basic approaches to generating cost estimates. The first one is a variant approach, in which cost estimates are based on variations of previous cost records. The other approach is the generative cost estimation, in which cost estimates are developed from scratch without taking previous cost records into consideration.

Optimistic and Pessimistic Cost Estimates. Using an adaptation of the Program Evaluation and Review Technique (PERT) formula, we can combine optimistic and pessimistic cost estimates. Let:

O = optimistic cost estimate
M = most likely cost estimate
P = pessimistic cost estimate

Then, the cost can be estimated as

$$E[C] = \frac{O + 4M + P}{6}$$

and the cost variance can be estimated as

$$V[C] = \left[\frac{P - O}{6}\right]^2$$

Budgeting and capital allocation

Budgeting involves sharing limited resources between several project groups or functions in a project environment. Budget analysis can serve any of the following purposes:

- A plan for resources expenditure
- A project selection criterion
- A projection of project policy
- A basis for project control

- A performance measure
- A standardization of resource allocation
- An incentive for improvement

Top-down budgeting

Top-down budgeting involves collecting data from upper-level sources such as top and middle managers. The numbers supplied by the managers may come from their personal judgment, past experience, or past data on similar project activities. The cost estimates are passed to lower-level managers, who then break the estimates down into specific work components within the project. These estimates may, in turn, be given to line managers, supervisors, and lead workers to continue the process until individual activity costs are obtained. Top management provides the global budget, while the functional level worker provides specific budget requirements for the project items.

Bottom-up budgeting

In this method, elemental activities, and their schedules, descriptions, and labor skill requirements are used to construct detailed budget requests. Line workers familiar with specific activities are requested to provide cost estimates. Estimates are made for each activity in terms of labor time, materials, and machine time. The estimates are then converted to an appropriate cost basis. The dollar estimates are combined into composite budgets at each successive level up the budgeting hierarchy. If estimate discrepancies develop, they can be resolved through the intervention of senior management, middle management, functional managers, project manager, accountants, or standard cost consultants.

Elemental budgets may be developed on the basis of the times progress of each part of the project. When all the individual estimates are gathered, we obtain a composite budget estimate. Analytical tools such as learning curve analysis, work sampling, and statistical estimation may be employed in the cost estimation and budgeting processes.

Mathematical formulation of capital allocation

Capital rationing involves selecting a combination of projects that will optimize the return on investment. A mathematical formulation of the capital budgeting problem is presented below:

$$\text{Maximize } z = \sum_{i=1}^{n} v_i x_i$$

$$\text{Subject to } \sum_{i=1}^{n} c_i x_i \le B$$

$$x_i = 0,1; \quad i = 1,\ldots,n$$

where:

n = number of projects

v_i = measure of performance for project i (e.g., present value)

c_i = cost of project i

x_i = indicator variable for project i

B = budget availability level

A solution of the above model will indicate which projects should be selected in combination with which projects. The example that follows illustrates a capital rationing problem.

Example

Planning of portfolio of projects is essential in resource limited projects. The capital rationing example presented here (Badiru and Pulat, 1995) involves the determination of the optimal combination of project investments so as to maximize total return on investment. Suppose that a project analyst is given N projects, X_1, X_2, X_3, . . ., X_N, with the requirement to determine the level of investment in each project so that total investment return is maximized subject to a specified limit on available budget. The projects are not mutually exclusive.

The investment in each project starts at a base level b_i ($i=1, 2, . . ., N$) and increases by variable increments k_{ij} ($j=1, 2, 3, . . ., K_i$), where K_i is the number of increments used for project i. Consequently, the level of investment in project X_i is defined as

$$x_i = b_i + \sum_{j=1}^{K_i} k_{ij}$$

where:

$$x_i \ge 0 \quad \forall i$$

For most cases, the base investment will be zero. In those cases, we will have $b_i = 0$. In the modeling procedure used for this problem, we have

$$X_i = \begin{cases} 1 & \text{if the investment in project } i \text{ is greater than zero} \\ 0 & \text{otherwise} \end{cases}$$

and

$$Y_{ij} = \begin{cases} 1 & \text{if } j\text{th increment of alternative } i \text{ is used} \\ 0 & \text{otherwise} \end{cases}$$

The variable x_i is the actual level of investment in project i, while X_i is an indicator variable indicating whether or not project i is one of the projects selected for investment. Similarly, k_{ij} is the actual magnitude of the jth increment while Y_{ij} is an indicator variable that indicates whether or not the jth increment is used for project i. The maximum possible investment in each project is defined as M_i such that

$$b_i \le x_i \le M_i$$

There is a specified limit, B, on the total budget available to invest such that

$$\sum_i x_i \le B$$

There is a known relationship between the level of investment, x_i, in each project and the expected return, $R(x_i)$. This relationship will be referred to as the *utility function, $f(\cdot)$,* for the project. The utility function may be developed through historical data, regression analysis, and forecasting models. For a given project, the utility function is used to determine the expected return, $R(x_i)$, for a specified level of investment in that project. That is,

$$R(x_i) = f(x_i)$$

$$= \sum_{j=1}^{K_i} r_{ij} Y_{ij}$$

where r_{ij} is the incremental return obtained when the investment in project i is increased by k_{ij}. If the incremental return decreases as the level of investment increases, the utility function will be *concave*. In that case, we will have the following relationship:

$$r_{ij} \ge r_{ij+1} \quad or \quad r_{ij} - r_{ij} + 1 \ge 0$$

Thus,

$$Y_{ij} \ge Y_{ij+1} \quad or \quad Y_{ij} - Y_{ij+1} \ge 0$$

so that only the first n increments ($j = 1, 2, \ldots, n$) that produce the highest returns are used for project i.

If the incremental returns do not define a concave function, $f(x_i)$, then one has to introduce the inequality constraints presented above into the optimization model. Otherwise, the inequality constraints may be left out of the model, since the first inequality, $Y_{ij} \geq Y_{ij+1}$, is always implicitly satisfied for concave functions. Our objective is to maximize the total return. That is,

$$\text{Maximize } Z = \sum_i \sum_j r_{ij} Y_{ij}$$

Subject to the following constraints:

$$x_i = b_i + \sum_j k_{ij} Y_{ij} \quad \forall i$$

$$b_i \leq x_i \leq M_i \quad \forall i$$

$$Y_{ij} \geq Y_{ij+1} \quad \forall i, j$$

$$\sum_i x_i \leq B$$

$$x_i \geq 0 \quad \forall i$$

$$Y_{ij} = 0 \text{ or } 1 \quad \forall i, j$$

Now suppose we are given four projects (i.e., $N = 4$) and a budget limit of $10 million.

For example, if an incremental investment of $0.20 million from stage 2 to stage 3 is made in project 1, the expected incremental return from the project will be $0.30 million. Thus, a total investment of $1.20 million in project 1 will yield a total return of $1.90 million.

The question addressed by the optimization model is to determine how many investment increments should be used for each project. That is, when should we stop increasing the investments in a given project? Obviously, for a single project, we would continue to invest as long as the incremental returns are larger than the incremental investments. However, for multiple projects, investment interactions complicate the decision so that investment in one project cannot be independent of the other projects. The LP model of the capital rationing example was solved with LINDO software. The solution indicates the following values for Y_{ij}.

Project 1:

$$Y11 = 1, \quad Y12 = 1, \quad Y13 = 1, \quad Y14 = 0, \quad Y15 = 0$$

Thus, the investment in project 1 is $X1 = \$1.20$ million. The corresponding return is $1.90 million.

Project 2:

$$Y21 = 1, \quad Y22 = 1, \quad Y23 = 1, \quad Y24 = 1, \quad Y25 = 0, \quad Y26 = 0, \quad Y27 = 0$$

Thus, the investment in project 2 is $X2 = \$3.80$ million. The corresponding return is $6.80 million.

Project 3:

$$Y31 = 1, \quad Y32 = 1, \quad Y33 = 1, \quad Y34 = 1, \quad Y35 = 0, \quad Y36 = 0, \quad Y37 = 0$$

Thus, the investment in project 3 is $X3 = \$2.60$ million. The corresponding return is $5.90 million.

Project 4:

$$Y41 = 1, \quad Y42 = 1, \quad Y43 = 1$$

Thus, the investment in project 4 is $X4 = \$2.35$ million. The corresponding return is $3.70 million.

The total investment in all four projects is $9,950,000. Thus, the optimal solution indicates that not all of the $10,000,000 available should be invested. The expected return from the total investment is $18,300,000. This translates into 83.92% return on investment.

The optimal solution indicates an unusually large return on total investment. In a practical setting, expectations may need to be scaled down to fit the realities of the project environment. Not all optimization results will be directly applicable to real situations. Possible extensions of the above model of capital rationing include the incorporations of risk and time value of money into the solution procedure. Risk analysis would be relevant, particularly for cases where the levels of returns for the various levels of investment are not known with certainty. The incorporation of time value of money would be useful if the investment analysis is to be performed for a given planning horizon. For example, we might need to make investment decisions to cover the next five years rather than just the current time.

Cost monitoring

As a project progresses, costs can be monitored and evaluated to identify areas of unacceptable cost performance. A plot of cost versus time for projected cost and actual cost can reveal a quick identification of when cost overruns occur in a project.

Plots similar to those presented above may be used to evaluate cost, schedule, and time performance of a project. An approach similar to the profit ratio presented earlier may be used together with the plot to evaluate the overall cost performance of a project over a specified planning horizon. Presented below is a formula for *cost performance index (CPI)*:

$$CPI = \frac{Area\,of\,cost\,benefit}{Area\,of\,cost\,benefit + Area\,of\,cost\,overrun}$$

As in the case of the profit ratio, CPI may be used to evaluate the relative performance of several project alternatives or to evaluate the feasibility and acceptability of an individual alternative.

Project balance technique

One other approach to monitoring cost performance is the project balance technique. The technique helps in assessing the economic state of a project at a desired point in time in the life cycle of the project. It calculates the net cash flow of a project up to a given point in time. The project balance is calculated as

$$B(i)_t = S_t - P(1+i)^t + \sum_{k=1}^{t} PW_{income}(i)_k$$

where:

$B(i)_i$ = project balance at time t at an interest rate of $i\%$ per period

$PW_{income}(i)_t$ = present worth of net income from the project up to time t

P = initial cost of the project

S_t = salvage value at time t

The project balanced at time t gives the net loss or net project associated with the project up to that time.

Cost control system

Contract management involves the process by which goods and services are acquired, utilized, monitored, and controlled in a project. Contract management addresses the contractual relationships from the initiation of a project to the completion of the project (i.e., completion of services and/ or handover of deliverables). Some of the important aspects of contract management are

- Principles of contract law
- Bidding process and evaluation
- Contract and procurement strategies
- Selection of source and contractors
- Negotiation
- Worker safety considerations
- Product liability
- Uncertainty and risk management
- Conflict resolution

In 1967, the US Department of Defense (DoD) introduced a set of 35 standards or criteria which contractors must comply with under cost or incentive contracts. Although it is no longer actively required by DoD, the concepts and techniques behind the criteria are still very much applicable and useful for effective management of project costs. Further, the criteria formed the basis of more recent cost-management techniques. The system of criteria is referred to as the *cost and schedule control systems criteria* (C/SCSC). Many government agencies now require compliance with C/SCSC for major contracts. The purpose is to manage the risk of cost overrun to the government. The system presents an integrated approach to cost and schedule management. It is now widely recognized and used in major project environments. It is intended to facilitate greater uniformity and provide advance warning about impending schedule or cost overruns.

The topics covered by C/SCSC include cost estimating and fore-casting, budgeting, cost control, cost reporting, earned value analysis, resource allocation and management, and schedule adjustments. The important link between all of these is the dynamism of the relationship between performance, time, and cost. This is essentially a multi-objective problem. Because performance, time, and cost objectives cannot be satisfied equally well, concessions or compromises would need to be worked out in implementing C/SCSC.

Another dimension of the performance–time–cost relationship is the US Air Force's R&M 2000 standard which addresses *reliability* and *maintainability* of systems. R&M 2000 is intended to integrate reliability and maintainability into the performance, cost, and schedule management for government contracts. C/SCSC and R&M 2000 constitute an effective guide for project design. To comply with C/SCSC, contractors must use standardized planning and control methods based on *earned value*. Earned value refers to the actual dollar value of work performed at a given point in time compared to planned cost for the work.

This is different from the conventional approach of measuring actual versus planned, which is explicitly forbidden by C/SCSC. In the conventional approach, it is possible to misrepresent the actual content (or value) of the work accomplished. The work-rate analysis technique presented in another section can be useful in overcoming the deficiencies of the conventional approach. C/SCSC is developed on a work content basis using the following factors:

- The actual cost of work performed (ACWP), which is determined on the basis of the data from cost accounting and information systems
- The budgeted cost of work scheduled (BCWS) or baseline cost determined by the costs of scheduled accomplishments
- The budgeted cost of work performed (BCWP) or earned value, the actual work of effort completed as of a specific point in time

The following equations can be used to calculate cost and schedule variances for work packages at any point in time.

$$\text{Cost variance} = \text{BCWP} - \text{ACWP}$$

$$\text{Percent cost variance} = (\text{Cost variance}/\text{BCWP})\,100$$

$$\text{Schedule variance} = \text{BCWP} - \text{BCWS}$$

$$\text{Percent schedule variance} = (\text{Schedule variance}/\text{BCWS})\,100$$

$$\text{ACWP and remaining funds} = \text{Target cost}\,(\text{TC})$$

$$\text{ACWP} + \text{cost to complete} = \text{Estimated cost at completion}\,(\text{EAC})$$

Sources of capital

Financing a project means raising capital for the project. Capital is a resource consisting of funds available to execute a project. Capital includes not only privately owned production facilities but also public investment. Public investments provide the infrastructure of the economy such as roads, bridges, water supply, and so on. Other public capital that indirectly supports production and private enterprise include schools, police stations, central financial institutions, and postal facilities.

If the physical infrastructure of the economy is lacking, the incentive for private entrepreneurs to invest in production facilities is likely to be lacking also. The government or community leaders can create the atmosphere for free enterprise by constructing better roads, providing better public safety, facilities, and encouraging ventures that assure adequate support services.

As far as project investment is concerned, what can be achieved with project capital is very important. The avenues for raising capital funds include banks, government loans or grants, business partners, cash reserves, and other financial institutions. The key to the success of the free enterprise system is the availability of capital funds and the availability of sources to invest the funds in ventures that yield products needed by the society. Some specific ways that funds can be made available for business investments are discussed below:

Commercial Loans. Commercial loans are the most common sources of project capital. Banks should be encouraged to loan money to entrepreneurs, particularly those just starting business. Government guarantees may be provided to make it easier for the enterprise to obtain the needed funds.

Bonds and Stocks. Bonds and stocks are also common sources of capital. National policies regarding the issuance of bonds and stocks can be developed to target specific project types to encourage entrepreneurs.

Interpersonal Loans. Interpersonal loans are unofficial means of raising capital. In some cases, there may be individuals with enough personal funds to provide personal loans to aspiring entrepreneurs. But presently, there is no official mechanism that handles the supervision of interpersonal business loans. If a supervisory body exists at a national level, wealthy citizens will be less apprehensive about loaning money to friends and relatives for business purposes. Thus, the wealthy citizens can become a strong source of business capital.

Foreign Investment. Foreign investments can be attracted for local enterprises through government incentives. The incentives may be in terms of attractive zoning permits, foreign exchange permits, or tax breaks.

Investment Banks. The operations of investment banks are often established to raise capital for specific projects. Investment banks buy securities from enterprises and resell them to other investors. Proceeds from these investments may serve as a source of business capital.

Mutual Funds. Mutual funds represent collective funds from a group of individuals. The collective funds are often large enough to provide capitals for business investments. Mutual funds may be established by individuals or under the sponsorship of a government agency. Encouragement and support should be provided for the group to spend the money for business investment purposes.

Supporting Resources. A clearing house of potential goods and services that a new project can provide may be established by the government. New entrepreneurs interested in providing the goods and services should be encouraged to start relevant enterprises. They should be given access to technical, financial, and information resources to facilitate starting production operations. As an example, the state of Oklahoma, under the auspices of the Oklahoma Center for the

Advancement of Science and Technology (OCAST), has established a resource database system. The system, named TRAC (Technical Resource Access Center), provides information about resources and services available to entrepreneurs in Oklahoma. The system is linked to the statewide economic development information system. This is a clearing house arrangement that will facilitate access to resources for project management.

The time value of money is an important factor in project planning and control. This is particularly crucial for long-term projects that are subject to changes in several cost parameters. Both the timing and quantity of cash flows are important for project management. The evaluation of a project alternative requires consideration of the initial investment, depreciation, taxes, inflation, economic life of the project, salvage value, and cash flows.

Activity-based costing

Activity-based costing (ABC) has emerged as an appealing costing technique in industry. The major motivation for ABC is that it offers an improved method to achieve enhancements in operational and strategic decisions. Activity-based costing offers a mechanism to allocate costs in direct proportion to the activities that are actually performed. This is an improvement over the traditional way of generically allocating costs to departments. It also improves the conventional approaches to allocating overhead costs.

The use of PERT/CPM, precedence diagramming, and the recently developed approach of critical resource diagramming (Badiru and Pulat, 1995) can facilitate task decomposition to provide information for activity-based costing. Some of the potential impacts of ABC on a production line are:

- Identification and removal of unnecessary costs
- Identification of the cost impact of adding specific attributes to a product
- Indication of the incremental cost of improved quality
- Identification of the value-added points in a production process
- Inclusion of specific inventory carrying costs
- Provision of a basis for comparing production alternatives
- Ability to assess "what-if" scenarios for specific tasks

Activity-based costing is just one component of the overall activity-based management in an organization. Activity-based management involves a more global management approach to planning and control of organizational endeavors. This requires consideration for product planning, resource allocation, productivity management, quality control, training, line balancing, value analysis, and a host of other organizational responsibilities. Thus, while activity-based costing is important, one must not lose sight of the universality of the environment in which it is expected to operate. And, frankly, there are some processes where functions are so intermingled that decomposition into specific activities may be difficult. Major considerations in the implementation of ABC are:

- Resources committed to developing activity-based information and cost
- Duration and level of effort needed to achieve ABC objectives
- Level of cost accuracy that can be achieved by ABC
- Ability to track activities based on ABC requirements
- Handling the volume of detailed information provided by ABC
- Sensitivity of the ABC system to changes in activity configuration

Income analysis can be enhanced by activity-based costing approach. Similarly, instead of allocating manufacturing overhead on the basis of direct labor costs, an activity-based costing analysis could be performed. The specific ABC cost components can be further broken down if needed. A spreadsheet analysis would indicate the impact on net profit as specific cost elements are manipulated.

Conclusion

In any manufacturing enterprise, computations and analyses similar to those illustrated in this chapter are crucial in controlling and enhancing the bottom-line survival of the organization. Analysts can adapt and extend the techniques presented in this chapter for application to the prevailing scenarios in their respective organizations.

References

Badiru, A. B. (2016), "Equity breakeven point: A graphical and tabulation tool for engineering managers," *Engineering Management Journal*, 28(4), 249–255.
Badiru, A. B. and O. A. Omitaomu (2007), *Computational Economic Analysis for Engineering and Industry*, Taylor & Francis Group/CRC Press, Boca Raton, FL.
Badiru, A. B. and P. S. Pulat (1995), *Comprehensive Project Management: Integrating Optimization Models, Management Principles, and Computers*, Prentice-Hall, Englewood Cliffs, NJ.

chapter four

Learning curves in manufacturing operations

Introduction

Manufacturing technology has several unique characteristics that make its management challenging. Some of these characteristics include frequent life cycle changes and uncertainty in the operating environment. Learning curve analysis offers a viable approach for evaluating manufacturing systems, where human learning and forgetting are involved. With an effective learning curve evaluation, an assessment can be made of how a manufacturing technology project meets organizational objectives and maximizes its benefits to the organization. Figure 4.1 illustrates a conventional curve of the life cycle of a manufacturing technology project as a function of resource infusion. In reality there will be dips in the curve due to the effects of learning and forgetting phenomena. It is essential to be able to predict locations of such dips so that an accurate assessment of the overall technology performance can be done. The half-life theory of learning curves introduced by Badiru and Ijaduola (2009) offers one good technique for an accurate assessment.

The fact is that the fast pace of technology affects learning curves and the frequent changes of technology degrades learning curves. Thus, specialized analytical assessment of learning curves is needed for manufacturing technology project management. The degradation of learning curves is often depicted analytically by incorporating forgetting components into conventional learning curves, as has been shown in the literature over the past few decades (Badiru, 1994, 1995a; Jaber and Sikstrom, 2004; Jaber et al., 2003; Jaber and Bonney, 1996; Jaber and Bonney, 2007; Nembhard and Osothsilp, 2001; Nembhard and Uzumeri, 2000; Sule, 1978; Globerson et al., 1998).

Half-life theory of learning curves

As another approach to capturing the essence of forgetting in learning curves, Badiru and Ijaduola (2009) introduced the half-life theory of learning curves. Traditionally, the standard time has been used as an indication of when learning should cease or when resources need to be transferred

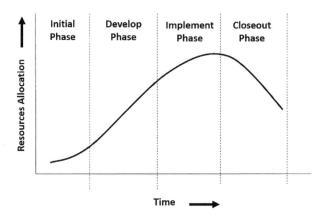

Figure 4.1 Manufacturing technology project lifecycle.

to another job. It is possible that half-life theory can supplement standard time analysis. The half-life approach will encourage researchers and practitioners to reexamine conventional applications of existing learning curve models. Organizations invest in people, work process, and technology for the purpose of achieving performance improvement. The systems nature of such investment strategy requires that the investment be strategically planned over multiple years. Thus, changes in learning curve profiles over those years become very crucial. Forgetting analysis and half-life computations can provide additional insights into learning curve changes. Through the application of robust learning curve analysis, system enhancement can be achieved in terms of cost, time, and performance with respect to strategic investment of funds and other organizational assets in people, process, and technology as shown in Figure 4.2. The predictive capability of learning curves is helpful in planning for integrated system performance improvement.

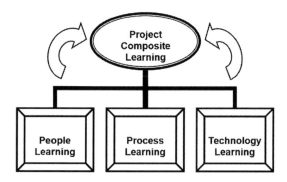

Figure 4.2 Learning components of people–process–technology integration.

Formal analysis of learning curves first emerged in the mid-1930s in connection with the analysis of the production of airplanes (Wright, 1936). Learning refers to the improved operational efficiency and cost reduction obtained from repetition of a task. This has a direct impact for training purposes and the design of work. Workers learn and improve by repeating operations. But they also regress due to the impact of forgetting, prolonged breaks, work interruption, and natural degradation of performance. Half-life computations can provide a better understanding of actual performance levels over time. Half-life is the amount of time it takes for a quantity to diminish to half of its original size through natural processes. Duality is of natural interest in many real-world processes. We often speak of "twice as much" and "half as much" as benchmarks for process analysis. In economic and financial principles, the "rule of 72" refers to the length of time required for an investment to double in value. These common "double" or "half" concepts provide the motivation for half-life analysis.

The usual application of half-life is in natural sciences. For example, in physics, the half-life is a measure of the stability of a radioactive substance. In practical terms, the half-life attribute of a substance is the time it takes for one-half of the atoms in an initial magnitude to disintegrate. The longer the half-life of a substance, the more stable it is. This provides a good analogy for modeling learning curves with the recognition of increasing performance or decreasing cost with respect to the passage of time. The approach provides another perspective to the large body of literature on learning curves. Badiru and Ijaduola (2009) present the following formal definitions:

> *For learning curves*: *Half-life* is the production level required to reduce cumulative average cost per unit to half of its original size.
> *For forgetting curves*: *Half-life* is the amount of time it takes for performance to decline to half its original magnitude.

Human-technology performance degradation

Although there is an extensive collection of classical studies of *improvement* due to learning curves, only very limited attention has been paid to performance *degradation* due to the impact of forgetting. Some of the classical works on process improvement due to learning include Belkaoui (1976), Camm et al. (1987), Liao (1979), Mazur and Hastie (1978), McIntyre (1977), Nanda (1979), Pegels (1976), Richardson (1978), Smith (1989), Smunt (1986), Sule (1978), Womer (1979, 1981, 1984), Womer and Gulledge (1983), Yelle (1976, 1979, 1983). It is only in recent years that the recognition of "forgetting" curves began to emerge, as can be seen in more recent literature (Badiru, 1995a; Jaber and Sikstrom, 2004; Jaber et al., 2003;

Jaber and Bonney, 2003, 2007; Jaber and Guiffrida, 2008). The new and emerging research on the forgetting components of learning curves provides the motivation for studying half-life properties of learning curves. Performance decay can occur due to several factors, including lack of training, reduced retention of skills, lapse in performance, extended breaks in practice, and natural forgetting. The conventional learning curve equation introduced by Wright (1936) has a drawback whereby the cost/time per unit approaches zero as the cumulative output approaches infinity. That is:

$$\lim_{x \to \infty} C(x) = \lim_{x \to \infty} C_1 x^{-b} \to 0$$

Researchers who initially embraced the Wright's learning curve (WLC) assumed a lower bound for the equation such that WLC could be represented as:

$$C(x) = \begin{cases} C_1 x^{-b}, & \text{if } x < x_s \\ C_s, & \text{otherwise} \end{cases},$$

where x_s is the number of units required to reach standard cost C_s. A half-life analysis can reveal more information about the properties of WLC, particularly when we consider the operating range of $x_0 < x_s$.

Half-life derivations

Learning curves present the relationship between cost (or time) and level of activity on the basis of the effect of learning. An early study by Wright (1936) disclosed the "80 percent learning" effect, which indicates that a given operation is subject to a 20% productivity improvement each time the activity level or production volume **doubles**. The proposed half-life approach is the antithesis of the double-level milestone. Learning curves can serve as a predictive tool for obtaining time estimates for tasks that are repeated within a project life cycle. A new learning curve does not necessarily commence each time a new operation is started, since workers can sometimes transfer previous skills to new operations. The point at which the learning curve begins to flatten depends on the degree of similarity of the new operation to previously performed operations. Typical learning rates that have been encountered in practice range from 70% to 95%. Several alternate models of learning curves have been presented in the literature, including *Log-linear model, S-curve model, Stanford-B model, DeJong's learning formula, Levy's adaptation function, Glover's learning formula, Pegels' exponential function, Knecht's upturn model,* and *Yelle's product model.* The basic log-linear model is the most popular learning curve model. It expresses a dependent variable (e.g., production cost) in terms of some independent variable (e.g., cumulative production). The model states that the improvement in productivity is constant (i.e., it has a constant slope) as output increases. That is:

$$C(x) = C_1 x^{-b}$$

where:
    $C(x)$ = cumulative average cost of producing x units
      C_1 = cost of the first unit
       x = cumulative production unit
      b = learning curve exponent.

The expression for $C(x)$ is practical only for $x > 0$. This makes sense because learning effect cannot realistically kick in until at least one unit ($x \geq 1$) has been produced. For the standard log-linear model, the expression for the learning rate, p, is derived by considering two production levels where one level is double the other. The performance curve, $P(x)$, can be defined as the reciprocal of the average cost curve, $C(x)$. Thus, we have:

$$P(x) = \frac{1}{C(x)},$$

which will have an increasing profile compared to the asymptotically declining cost curve. In terms of practical application, learning to drive is one example where maximum performance can be achieved in relatively short time compared to the half-life of performance. That is, learning is steep, but the performance curve is relatively flat after a steady state is achieved. The application of half-life analysis to learning curves can help address questions such as the ones below:

- How fast and how far can system performance be improved?
- What are the limitations to system performance improvement?
- How resilient is a system to shocks and interruptions to its operation?
- Are the performance goals that are set for the system achievable?

Half-life of the log-linear model

Figure 4.3 shows the basic log-linear model, with the half-life point indicated as $x_{1/2}$. The half-life of the log-linear model is computed as follows: Let:

    C_0 = initial performance level
    $C_{1/2}$ = performance level at half-life

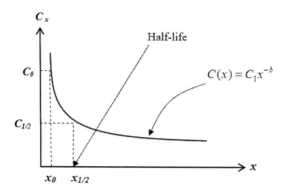

Figure 4.3 Profile of basic learning curve with half-life point.

$$C_0 = C_1 x_0^{-b} \quad \text{and} \quad C_{1/2} = C_1 x_{1/2}^{-b}$$

But $C_{1/2} = \dfrac{1}{2} C_0$

Therefore, $C_1 x_{1/2}^{-b} = \dfrac{1}{2} C_1 x_0^{-b}$, which leads to $x_{1/2}^{-b} = \dfrac{1}{2} x_0^{-b}$, which, by taking the $(-1/b)^{\text{th}}$ exponent of both sides, simplifies to yield the following expression as the general expression for the standard log-linear learning curve model,

$$x_{1/2} = \left(\frac{1}{2}\right)^{-\frac{1}{b}} x_0, \quad x_0 \geq 1$$

where $x_{1/2}$ is the half-life and x_0 is the initial point of operation. We refer to $x_{1/2}$ (Figure 4.3) as the *First-Order half-life*. The *Second-Order half-life* is computed as the time corresponding to half of the preceding half. That is:

$$C_1 x_{1/2(2)}^{-b} = \frac{1}{4} C_1 x_0^{-b},$$

which simplifies to yield:

$$x_{1/2(2)} = (1/2)^{-2/b} x_0,$$

Similarly, the *Third-Order half-life* is derived to obtain:

$$x_{1/2(3)} = (1/2)^{-3/b} x_0,$$

In general, the kth-*Order half-life* for the log-linear model is represented as:

$$x_{1/2(k)} = (1/2)^{-k/b} x_0,$$

Half-life computational examples

This section uses examples of log-linear learning curves with b=0.75 and b=0.3032 respectively to illustrate the characteristics of learning which can dictate the half-life behavior of the overall learning process. Knowing the point where the half-life of each curve occurs can be very useful in assessing learning retention for the purpose of designing training programs or designing work. For $C(x) = 250x^{-0.75}$, the First-Order half-life is computed as:

$$x_{1/2} = (1/2)^{-1/0.75} x_0, \quad x_0 \geq 1$$

If the above expression is evaluated for $x_0=2$, the First-Order half-life yields $x_{1/2}=5.0397$; which indicates a fast drop in the value of $C(x)$. $C(2) =148.6509$ corresponding to a half-life of 5.0397. Note that $C(5.0397) = 74.7674$, which is about half of 148.6509. The conclusion from this analysis is that if we are operating at the point $x=2$, we can expect the curve to reach its half-life decline point at $x=5$. For $C(x) = 240.03x^{-0.3032}$, the First-Order half-life is computed as:

$$x_{1/2} = (1/2)^{-1/0.3032} x_0, \quad x_0 \geq 1$$

If we evaluate the above function for $x_0=2$; the First-Order half-life is $x_{1/2} = 19.6731$. Several models and variations of learning curves are used in practice. Models are developed through one of the following approaches:

1. Conceptual models
2. Theoretical models
3. Observational models
4. Experimental models
5. Empirical models

The S-curve model

The S-curve (Towill and Kaloo, 1978) is based on an assumption of a gradual start-up. The function has the shape of the cumulative normal distribution function for the start-up curve and the shape of an operating characteristics function for the learning curve. The gradual start-up is based on the fact that the early stages of production are typically in a transient state with changes in tooling, methods, materials, design, and even changes in the work force. The basic form of the S-curve function is:

$$C(x) = C_1 + M(x+B)^{-b}$$

$$MC(x) = C_1 \left[M + (1-M)(x+B)^{-b} \right]$$

where:

    $C(x)$ = learning curve expression
     b = learning curve exponent
    $M(x)$ = marginal cost expression
      C_1 = cost of first unit
      M = incompressibility factor (a constant)
      B = equivalent experience units (a constant).

Assumptions about at least three out of the four parameters $(M, B, C_1,$ and $b)$ are needed to solve for the fourth one. Using the $C(x)$ expression and derivation procedure outlined earlier for the log-linear model, the half-life equation for the S-curve learning model is derived to be:

$$x_{1/2} = (1/2)^{-1/b} \left[\frac{M(x_0 + B)^{-b} - C_1}{M} \right]^{-1/b} - B$$

where:

    $x_{1/2}$ = half-life expression for the S-curve learning model
    x_0 = initial point of evaluation of performance on the learning curve

In terms of practical application of the S-curve, consider when a worker begins learning a new task. The individual is slow initially at the tail end of the S-curve. But the rate of learning increases as time goes on, with additional repetitions. This helps the worker to climb the steep-slope segment of the S-curve very rapidly. At the top of the slope, the worker is classified as being proficient with the learned task. From then on, even if the worker puts much effort into improving upon the task, the resultant learning will not be proportional to the effort expended. The top end of the S-curve is often called the slope of *diminishing returns.* At the top of the S-curve, workers succumb to the effects of *forgetting* and other performance impeding factors. As the work environment continues to change, a worker's level of skill and expertise can become obsolete. This is an excellent reason for the application of half-life computations.

The Stanford-B model

The Stanford-B model is represented as:

$$UC(x) = C_1 (x+B)^{-b}$$

where:
$UC(x)$ = direct cost of producing the xth unit
b = learning curve exponent
C_1 = cost of the first unit when $B = 0$
B = slope of the asymptote for the curve

B = constant $(1 < B < 10)$. This is equivalent units of previous experience at the start of the process, which represents the number of units produced prior to first unit acceptance. It is noted that when $B = 0$, the Stanford-B model reduces to the conventional log-linear model. The general expression for the half-life of the Stanford-B model is derived to be:

$$x_{1/2} = (1/2)^{-1/b}(x_0 + B) - B$$

where:
$x_{1/2}$ = half-life expression for the Stanford-B learning model
x_0 = initial point of evaluation of performance on the learning curve

Badiru's multi-factor model

Badiru (1994) presents applications of learning and forgetting curves to productivity and performance analysis. One example presented used production data to develop a predictive model of production throughput. Two data replicates are used for each of ten selected combinations of cost and time values. Observations were recorded for the number of units representing double production levels. The resulting model has the functional form below and the graphical profile shown in Figure 4.4.

$$C(x) = 298.88x_1^{-0.31}x_2^{-0.13}$$

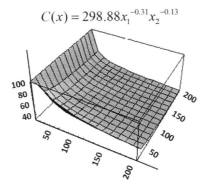

Figure 4.4 Bivariate model of learning curve.

where:

    $C(x)$ = cumulative production volume

      x_1 = cumulative units of Factor 1

      x_2 = cumulative units of Factor 2

      b_1 = first learning curve exponent = −0.31

      b_2 = second learning curve exponent = −0.13

A general form of the modeled multi-factor learning curve model is:

$$C(x) = C_1 x_1^{-b_1} x_2^{-b_2}$$

and the half-life expression for the multi-factor learning curve was derived to be:

$$x_{1(1/2)} = (1/2)^{-1/b_1} \left[\frac{x_{1(0)} x_{2(0)}^{b_2/b_1}}{x_{2(1/2)}^{b_2/b_1}} \right]^{-1/b_1}$$

$$x_{2(1/2)} = (1/2)^{-1/b_2} \left[\frac{x_{2(0)} x_{1(0)}^{b_1/b_2}}{x_{1(1/2)}^{b_2/b_1}} \right]^{-1/b_2}$$

where:

    $x_{i(1/2)}$ = half-life component due to Factor i (i = 1, 2)

    $x_{i(0)}$ = initial point of Factor i (i = 1, 2) along the multi-factor learning curve

    Knowledge of the value of one factor is needed to evaluate the other factor. Just as in the case of single-factor models, the half-life analysis of the multi-factor model can be used to predict when the performance metric will reach half of a starting value.

DeJong's learning formula

DeJong's learning formula is a power function which incorporates parameters for the proportion of manual activity in a task. When operations are controlled by manual tasks, the time will be compressible as successive units are completed. If, by contrast, machine cycle times control operations, then the time will be less compressible as the number of units increases. DeJong's formula introduces as an incompressible factor, M, into the log-linear model to account for the man-machine ratio. The model is expressed as:

$$C(x) = C_1 + Mx^{-b}$$

$$MC(x) = C_1 \left[M + (1 - M) x^{-b} \right]$$

where:
  $C(x)$ = learning curve expression
  $M(x)$ = marginal cost expression
  b = learning curve exponent
  C_1 = cost of first unit
  M = incompressibility factor (a constant)

When $M = 0$, the model reduces to the log-linear model, which implies a completely manual operation. In completely machine-dominated operations, $M = 1$. In that case, the unit cost reduces to a constant equal to C_1, which suggests that no learning-based cost improvement is possible in machine-controlled operations. This represents a condition of high incompressibility. This profile suggests impracticality at higher values of production. Learning is very steep and average cumulative production cost drops rapidly. The horizontal asymptote for the profile is below the lower bound on the average cost axis, suggesting an infeasible operating region as production volume gets high. The analysis above agrees with the fact that no significant published data is available on whether or not DeJong's learning formula has been successfully used to account for the degree of automation in any given operation. Using the expression, $MC(x)$, the marginal cost half-life of the DeJong's learning model is derived to be:

$$x_{1/2} = (1/2)^{-1/b} \left[\frac{(1 - M)x_0^{-b} - M}{2(1 - M)} \right]^{-1/b}$$

where:
  $x_{1/2}$ = half-life expression for the DeJong's learning curve marginal cost model
  x_0 = initial point of evaluation of performance on the marginal cost curve

If the $C(x)$ model is used to derive the half-life, then we obtain the following derivation:

$$x_{1/2} = (1/2)^{-1/b} \left[\frac{Mx_0^{-b} - C_1}{M} \right]^{-1/b}$$

where:
  $x_{1/2}$ = half-life expression for the DeJong's learning curve model
  x_0 = initial point of evaluation of performance on the DeJong's learning curve

Levy's technology adaptation function.

Recognizing that the log-linear model does not account for leveling off of production rate and the factors that may influence learning, Levy (1965) presented the following learning cost function:

$$MC(x) = \left[\frac{1}{\beta} - \left(\frac{1}{\beta} - \frac{x^{-b}}{C_1} \right) k^{-kx} \right]^{-1}$$

where:
    β = production index for the first unit;
    k = constant used to flatten the learning curve for large values of x.

The flattening constant, k, forces the curve to reach a plateau instead of continuing to decrease or turning in the upward direction. The half-life expression for Levy's learning model is a complex non-linear expression derived as shown below:

$$(1/\beta - x_{1/2}^{-b}/C_1)k^{-kx_{1/2}} = 1/\beta - 2[1/\beta - (1/\beta - x_0^{-b}/C_1)k^{-kx_0}]$$

where:
    $x_{1/2}$ = half-life expression for the Levy's learning curve model
    x_0 = initial point of evaluation of performance on the Levy's learning curve.

Knowledge of some of the parameters of the model is needed to solve for the half-life as a closed form expression.

Glover's learning model

Glover's learning formula (Glover, 1966) is a learning curve model that incorporates a work commencement factor. The model is based on a bottom-up approach which uses individual worker learning results as the basis for plant-wide learning curve standards. The functional form of the model is expressed as:

$$\sum_{i=1}^{n} y_i + a = C_1 \left(\sum_{i=1}^{n} x_i \right)^m$$

where:
    y_i = elapsed time or cumulative quantity;
    x_i = cumulative quantity or elapsed time;
    a = commencement factor;
    n = index of the curve (usually 1+b);
    m = model parameter.

This is a complex expression for which half-life expression is not easily computable. We defer the half-life analysis of Levy's learning curve model for further research by interested readers.

Pegels' exponential function

Pegels (1976) presented an alternate algebraic function for the learning curve. His model, a form of an exponential function of marginal cost, is represented as:

$$MC(x) = \alpha a^{x-1} + \beta$$

where α, β, and a are parameters based on empirical data analysis. The total cost of producing x units is derived from the marginal cost as follows:

$$TC(x) = \int \left(\alpha a^{x-1} + \beta \right) dx = \frac{\alpha a^{x-1}}{\ln(a)} + \beta x + c$$

where c is a constant to be derived after the other parameters are found. The constant can be found by letting the marginal cost, total cost, and average cost of the first unit to be all equal. That is, $MC_1 = TC_1 = AC_1$, which yields:

$$c = \alpha - \frac{\alpha}{\ln(a)}$$

The model assumes that the marginal cost of the first unit is known. Thus,

$$MC_1 = \alpha + \beta = y_0$$

Mathematical expression for the total labor cost in Pegels' start-up curves is expressed as:

$$TC(x) = \frac{a}{1-b} x^{1-b}$$

where:
$\quad$ x = cumulative number of units produced;
$\quad$ a, b = empirically determined parameters.

The expressions for marginal cost, average cost, and unit cost can be derived as shown earlier for other models. Using the total cost expression, TC(x), we derive the expression for the half-life of Pegels' learning curve model to be as shown below:

$$x_{1/2} = \left(\tfrac{1}{2} \right)^{-1/(1-b)} x_0$$

Knecht's upturn model

Knecht (1974) presents a modification to the functional form of the learning curve to analytically express the observed divergence of actual costs from those predicted by learning curve theory when units produced exceed 200. This permits the consideration of non-constant slopes for the learning curve model. If UC_x is defined as the unit cost of the xth unit, then it approaches 0 asymptotically as x increases. To avoid a zero limit unit cost, the basic functional form is modified. In the continuous case, the formula for cumulative average costs is derived as:

$$C(x) = \int_0^x C_1 z^b dz = \frac{C_1 x^{b+1}}{(1+b)}$$

This cumulative cost also approaches zero as x goes to infinity. Knecht alters the expression for the cumulative curve to allow for an upturn in the learning curve at large cumulative production levels. He suggested the functional form below:

$$C(x) = C_1 x^{-b} e^{cx}$$

where c is a second constant. Differentiating the modified cumulative average cost expression gives the unit cost of the xth unit as shown below. Figure 4.5 shows the cumulative average cost plot of Knecht's upturn function for values of $C_1 = 250$, $b = 0.25$, and $c = 0.25$.

$$UC(x) = \frac{d}{dx}\left[C_1 x^{-b} e^{cx} \right] = C_1 x^{-b} e^{cx} \left(c + \frac{-b}{x} \right).$$

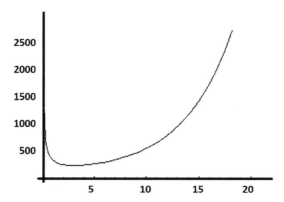

Figure 4.5 Knecht's cumulative average cost function for $c_1 = 250$, $b = 0.25$, and $c = 0.25$.

The half-life expression for Knecht's learning model turns out to be a non-linear complex function as shown below:

$$x_{1/2}e^{-cx_{1/2}/b} = (\tfrac{1}{2})^{-1/b}e^{-cx_0/b}x_0$$

where:

$x_{1/2}$ = half-life expression for the Knecht's learning curve model

x_0 = initial point of evaluation of performance on the Knecht's learning curve

Given that x_0 is known, iterative, interpolation, or numerical methods may be needed to solve for the half-life value.

Yelle's combined technology learning curve

Yelle (1979) proposed a learning curve model for products by aggregating and extrapolating the individual learning curve of the operations making up a product on a log-linear plot. The model is expressed as shown below:

$$C(x) = k_1 x_1^{-b_1} + k_2 x_2^{-b_2} + \cdots + k_n x_n^{-b_n}$$

where:

$C(x)$ = cost of producing the xth unit of the product

n = number of operations making up the product

$k_i x_i^{-b_i}$ = learning curve for the ith operation

Aggregated learning curves

In comparing the models discussed in the preceding sections, the deficiency of Knecht's model is that a product-specific learning curve seems to be a more reasonable model than an integrated product curve. For example, an aggregated learning curve with 96.6% learning rate obtained from individual learning curves with the respective learning rates of 80%, 70%, 85%, 80%, and 85% does not appear to represent reality. If this type of composite improvement is possible, then one can always improve the learning rate for any operation by decomposing it into smaller integrated operations. The additive and multiplicative approaches of reliability functions support the conclusion of the impracticality of Knecht's integrated model. Jaber and Guiffrida (2004) presented an aggregated form of the WLC where some of the items produced are defective and require reworking. The quality learning curve that they provide is of the form:

$$t(x) = y_1 x^{-b} + 2r_1 \left(\frac{p}{2}\right)^{1-\varepsilon} x^{1-2\varepsilon}$$

Where y_1 is the time to produce the first unit, r_1 is the time to rework the first defective unit, p is the probability of the process to go out-of-control ($p \ll 1$), and b is the learning exponent of the reworks learning curve. The variable $t(x)$ has three behavioral patterns, for $0 < b < \frac{1}{2}$ (Case I), $b = \frac{1}{2}$ (Case II), and $\frac{1}{2} < b < 1$ (Case III). Assuming no production error, we computed the half-life for $t(x)$ for case I as:

$$\textbf{Case I: } x_{1/2} = \left(\frac{1}{2}\right)^{-\frac{1}{b}} x \quad \text{and} \quad x_{1/2} = \left(\frac{1}{2}\right)^{-\frac{1}{1-2\varepsilon}} x$$

$$\textbf{Case II: } t(x) = y_1 x^{-b} + 2r_1 \left(\frac{\rho}{2}\right)^{1-\varepsilon} x^{1-2\varepsilon} \text{ reduces to } t(x) = y_1 x^{-b} + t(x) =$$

$y_1 x^{-b} + 2r_1 \sqrt{\dfrac{\rho}{2}}$, where $2r_1 \sqrt{\dfrac{\rho}{2}}$ is the lower bound, or the plateau of the learning curve.

Case III: The behavior of $t(x)$ follows that of the WLC: monotonically decreasing as cumulative output increases. It is noted that Jaber and Guiffrida (2008) assumed that the percentage defective reduces as the number of interruptions to restore the process increases. They found that $t(x)$ could converge to the WLC as the learning curve exponent, b, becomes insignificant.

Half-life of decline curves

Over the years, the decline curve technique has been extensively used by the oil and gas industry to evaluate future oil and gas predictions. These predictions are used as the basis for economic analysis to support development, property sale or purchase, industrial loan provisions, and also to determine if a secondary recovery project should be carried out. It is expected that the profile of a hyperbolic decline curve can be adapted for application to learning curve analysis. The graphical solution of the hyperbolic equation is through the use of a log-log paper which sometimes provides a straight line that can be extrapolated for a useful length of time to predict future production levels. This technique, however, sometimes failed to produce the straight line needed for extrapolation for some production scenarios. Furthermore, the graphical method usually involves some manipulation of data, such as shifting, correcting and/or adjusting scales, which eventually introduce bias into the actual data. In order to avoid the noted graphical problems of hyperbolic decline curves and to accurately predict future performance of producing well, a non-linear least-squares technique is often considered. This method does not require any straight-line extrapolation for future predictions. The mathematical analysis proceeds as follows: The general hyperbolic decline equation for oil production rate (q) as a function of time (t) can be represented as

$$q(t) = q_0 (1 + mD_0t)^{-1/m}$$

$$0 < m < 1$$

where:

$q(t)$ = oil production at time t

q_0 = initial oil production

D_0 = initial decline

m = decline exponent

Also, the cumulative oil production at time t, $Q(t)$ can be written as

$$Q(t) = \frac{q_0}{(m-1)D_0}\left[(1+mD_0t)^{\frac{m-1}{m}} - 1\right]$$

Where $Q(t)$ = cumulative production as of time t. By combing the above equations and performing some algebraic manipulations, it can be shown that

$$q(t)^{1-m} = q_0^{1-m} + (m-1)D_0q_0^{-m}Q(t),$$

which shows that the production at time t is a non-linear function of its cumulative production level. By rewriting the equations in terms of cumulative production, we have

$$Q(t) = \frac{q_0}{(1-m)D_0} + q(t)^{1-m}\frac{q_0^m}{(m-1)D_0}$$

It can be seen that the model can be investigated both in terms of conventional learning curve techniques, forgetting decline curve, and half-life analysis in a procedure similar to techniques presented earlier in this chapter. The forgetting function has the same basic form as the standard learning curve model, except that the forgetting rate will be negative, indicating a decay process. Figure 4.6 shows some of the possible profiles of the forgetting curve. Profile (a) shows a case where forgetting occurs rapidly along a convex curve. Profile (b) shows a case where forgetting occurs more slowly along a concave curve. Profile (c) shows a case where the rate of forgetting shifts from convex to concave along an S-curve.

The profile of the forgetting curve and its mode of occurrence can influence the half-life measure. This is further evidence that the computation of half-life can help distinguish between learning curves, particularly if a forgetting component is involved. The combination of the learning

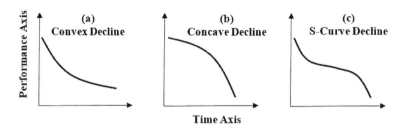

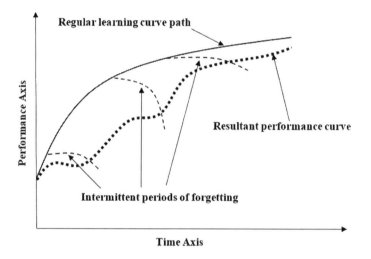

Figure 4.6 Profiles of learning decline curves.

Figure 4.7 Resolution of learn-forget performance curves.

and forgetting functions presents a more realistic picture of what actually occurs in a learning process. The combination is not necessarily as simple as resolving two curves to obtain a resultant curve. The resolution may particularly be complex in the case of intermittent periods of forgetting. Figure 4.7 shows representations of periods where forgetting occurs and the resulting learn-forget profile.

Conclusion

Degradation of performance occurs naturally either due to internal processes or externally imposed events, such as extended production breaks. For productivity assessment purposes, it may be of interest to determine the length of time it takes a production metric to decay to half of its original magnitude. For example, for career planning strategy, one may be interested in how long it takes for skills sets to degrade by half in relation to current technological needs of the workplace. The half-life phenomenon may be due to intrinsic factors, such as forgetting, or due to external factors, such as a shift in labor requirements. Half-life analysis can have application in intervention programs designed to achieve reinforcement of learning. It can also have application for assessing the sustainability of skills acquired through training programs. Further research on the theory of half-life of learning curves should be directed to topics such as the following:

- Half-life interpretations
- Training and learning reinforcement program
- Forgetting intervention and sustainability programs

In addition to the predictive benefits of half-life expressions, they also reveal the ad-hoc nature of some of the classical learning curve models that have been presented in the literature. We recommend that future efforts to develop learning curve models should also attempt to develop the corresponding half-life expressions to provide full operating characteristics of the models. Readers are encouraged to explore half-life analysis of other learning curve models not covered in this chapter. In some cases, a lower bound is incorporated into the conventional Wright's learning curve (WLC) such that WLC could be represented as:

$$C(x) = \begin{cases} C_1 x^{-b}, & \text{if } x < x_s \\ C_s, & \text{otherwise} \end{cases},$$

where x_s is the number of units required to reach standard cost C_s. Now, if we assume that for some $x_0 < x_s$, the half-life expression becomes:

$$x_{1/2} = \left(\frac{1}{2}\right)^{-\frac{1}{b}} x_0 > x_s.$$

What would this mean in an operational context particularly in dynamic science, technology, and engineering applications? Much research centered on life data needs to be done in this area. The half-life theory approach opens the door to many similar learning curve research inquiries.

References

Badiru, A. B. and A. Ijaduola (2009). "Half-life theory of learning curves for system performance analysis," *IEEE Systems Journal*, 3(2), 154–165.

Badiru, A. B. (1994). "Multifactor learning and forgetting models for productivity and performance analysis," *International Journal of Human Factors in Manufacturing*, 4(1), 37–54.

Badiru, A. B. (1995a). "Multivariate analysis of the effect of learning and forgetting on product quality," *International Journal of Production Research*, 33(3), 777–794.

Belkaoui, A. (1976). "Costing through learning," *Cost and Management*, 50(3), 36–40.

Camm, J. D., J. R. Evans and N. K. Womer (1987). "The unit learning curve approximation of total cost," *Computers and Industrial Engineering*, 12(3), 205–213.

Globerson, S., A. Nahumi, and S. Ellis (1998). "Rate of forgetting for motor and cognitive tasks," *International Journal of Cognitive Ergonomics*, 2, 181–191.

Glover, J. H. (1966). "Manufacturing progress functions: An alternative model and its comparison with existing functions," *International Journal of Production Research*, 4(4), 279–300.

Jaber, M. Y. and M. Bonney (1996). "Production breaks and the learning curve: *The forgetting phenomena*," *Applied Mathematical Modelling*, 20, 162–169.

Jaber, M. Y. and M. Bonney (2003). "Lot sizing with learning and forgetting in setups and in product quality," *International Journal of Production Economics*, 83(1), 95–111.

Jaber, M. Y. and M. Bonney (2007). "Economic manufacture quantity (emq) model with lot size dependent learning and forgetting rates," *International Journal of Production Economics*, 108(1–2), 359–367.

Jaber, M. Y. and A. L. Guiffrida (2004). "Learning curves for processes generating defects requiring reworks," *European Journal of Operational Research*, 159(3), 663–672.

Jaber, M. Y. and A. L. Guiffrida (2008). "Learning curves for imperfect production processes with reworks and process restoration interruptions," *European Journal of Operational Research*, 189(1), 93–104.

Jaber, M. Y. and S. Sikstrom (2004). "A numerical comparison of three potential learning and forgetting models," *International Journal of Production Economics*, 92(3), 281–294.

Jaber, M. Y., H. V. Kher, and D. Davis (2003). "Countering forgetting through training and deployment," *International Journal of Production Economics*, 85(1), 33–46.

Knecht, G. R. (1974). "Costing, technological growth, and generalized learning curves," *Operations Research Quarterly*, 25(3), 487–491.

Levy, F. K. (1965). "Adaptation in the production process," *Management Science*, 11(6), B136–B154.

Liao, W. M. (1979). "Effects of learning on resource allocation decisions," *Decision Sciences*, 10, 116–125.

Mazur, J. E. and R. Hastie (1978). "Learning as accumulation: A reexamination of the learning curve," *Psychological Bulletin*, 85, 1256–1274.

McIntyre, E. V. (1977). "Cost-volume-profit analysis adjusted for learning," *Management Science*, 24(2), 149–160.

Nanda, R. (1979). "Using learning curves in integration of production resources," *Proceedings of 1979 IIE Fall Conference*, 376–380.

Nembhard, D. A. and N. Osothsilp (2001). "An empirical comparison of forgetting models," *IEEE Transactions on Engineering Management*, 48, 283–291.

Nembhard, D. A. and M. V. Uzumeri (2000). "Experiential learning and forgetting for manual and cognitive tasks," *International Journal of Industrial Ergonomics*, 25, 315–326.

Pegels, Carl C. (1976). "Start up or learning curves – Some new approaches," *Decision Sciences*, 7(4), 705–713.

Richardson, W. J. (1978). "Use of learning curves to set goals and monitor progress in cost reduction programs," *Proceedings of 1978 IIE Spring Conference*, 235–239.

Smith, J. (1989). *Learning Curve for Cost Control*. Industrial Engineering and Management Press, Norcross, GA.

Smunt, T. L. (1986). "A comparison of learning curve analysis and moving average ratio analysis for detailed operational planning," *Decision Sciences*, 17, 39–53.

Sule, D. R. (1978). "The effect of alternate periods of learning and forgetting on economic manufacturing quantity," *AIIE Transactions*, 10(3), 338–343.

Towill, D. R. and U. Kaloo (1978). "Productivity drift in extended learning curves," *Omega*, 6(4), 295–304.

Womer, N. K. (1979). "Learning curves, production rate, and program costs," *Management Science*, 25(4), 312–219.

Womer, N. K. (1981). "Some propositions on cost functions," *Southern Economic Journal*, 47, 1111–1119.

Womer, N. K. (1984). "Estimating learning curves from aggregate monthly data," *Management Science*, 30(8), 982–992.

Womer, N. K. and T. R. Gulledge, Jr. (1983). "A dynamic cost function for an airframe production program," *Engineering Costs and Production Economics*, 7, 213–227.

Wright, T. P. (1936). "Factors affecting the cost of airplanes," *Journal of Aeronautical Science*, 3(2), 122–128.

Yelle, L. E. (1983). "Adding life cycles to learning curves," *Long Range Planning*, 16(6), 82–87.

Yelle, L. E. (1979). "The learning curve: Historical review and comprehensive survey," *Decision Sciences*, 10(2), 302–328.

Yelle, L. E. (1976). "Estimating learning curves for potential products," *Industrial Marketing Management*, 5(2/3), 147–154.

Bibliography

Alter, S. (1999). *Information Systems: A Management Perspective*. Third Edition. Addison Wesley Longman, Reading, MA.

Alter, S. (2002). *Information Systems: Foundation of E-Business*. Fourth Edition. Prentice Hall, Upper Saddle River, NJ.

Anderlohr, G. (1969). "What production breaks cost," *Industrial Engineering*, 20, 34–36.

Badiru, A. B. (2009). *STEP Project Management: Guide for Science, Technology, and Engineering Projects*. Taylor & Francis Group/CRC Press, Boca Raton, FL.

Badiru, A. B. (2008). *Triple C Model of Project Management:Communication, Cooperation, and Coordination*. Taylor & Francis Group/CRC Press, Boca Raton, FL.

Badiru, A. B. (1992). "Computational survey of univariate and multivariate learning curve models," *IEEE Transactions on Engineering Management*, 39(2), 176–188.

Badiru, A. B. (1995b). "Incorporating learning curve effects into critical resource diagramming," *Project Management Journal*, 26(2), 38–45.

Belkaoui, A (1986). *The Learning Curve*. Quorum Books, Westport, CT.

Ewusi-Mensah, K. (1997). "Critical issues in abandoned information systems development projects," *Communications of the ACM*, 40(9), 74–80.

McKenna, S. P. and A. I. Glendon (1985). "Occupational First Aid Training: Decay in Cardiopulmonary Resuscitation (CPR) Skills," *Journal of Occupational Psychology*, 58, 109–117.

Newnan, D. G., T. G. Eschenbach, and J. P. Lavelle (2004). *Engineering Economic Analysis*. Oxford University Press, New York.

Ross, J. W. and C. M. Beath (2002). Beyond the business case: New approaches to IT investments, *MIT Sloan Management Review*, Winter Edition, 51–59.

Sikstrom, S. and M. Y. Jaber (2002). "The power integration diffusion (PID) model for production breaks," *Journal of Experimental Psychology: Applied*, 8, 118–126.

Towill, D. R. and J. E. Cherrington, (1994). "Learning curve models for predicting the performance of advanced manufacturing technology," *International Journal of Advanced Manufacturing Technology*, 9(3), 195–203.

chapter five

Computational tools for manufacturing

Learning curve computations

There are several models of learning curves in use in business and industry. The log-linear model is perhaps the most extensively used. The log-linear model states that the improvement in productivity is constant (i.e., it has a constant slope) as output increases. There are two basic forms of the log-linear model: the average cost function and the unit cost function (see Badiru, 2012; Badiru et al., 2008; Badiru and Omitaomu, 2007, 2011; Heragu, 1997).

The average cost model

The average cost model is used more than the unit cost model. It specifies the relationship between the cumulative average cost per unit and cumulative production. The relationship indicates that cumulative cost per unit will decrease by a constant percentage as the cumulative production volume doubles. The *average cost* model is expressed as:

$$A_x = C_1 x^b$$

where:
A_x = cumulative average cost of producing x units
C_1 = cost of the first unit
x = cumulative production count
b = the learning curve exponent (i.e., constant slope of on log-log paper).

The relationship between the learning curve exponent, b, and the learning rate percentage, p, is given by:

$$b = \frac{\log p}{\log 2} \quad \text{or } p = 2^b$$

The derivation of the above relationship can be seen by considering two production levels where one level is double the other, as shown below.
Let Level I = x_1 and Level II = x_2 = $2x_1$. Then,

$$A_{x_1} = C_1(x_1)^b \quad \text{and } A_{x_2} = C_1(2x_1)^b$$

The percent productivity gain is then computed as:

$$p = \frac{C_1(2x_1)^b}{C_1(x_1)^b} = 2^b$$

On log-log paper, the model is represented by the following straight-line equation:

$$\log A_x = \log C_1 + b \log x,$$

where b is the constant slope of the line. It is from this straight line that the name log-linear was derived.

Computational example

Assume that 50 units of an item are produced at a cumulative average cost of \$20 per unit. Suppose we want to compute the learning percentage when 100 units are produced at a cumulative average cost of \$15 per unit. The learning curve analysis would proceed as follows:

Initial production level = 50 units; Average Cost = \$20

Double production level = 100 units; Cumulative Average Cost = \$15.

Using the log relationship, we obtain the following equations:

$$\log 20 = \log C_1 + b \log 50$$

$$\log 15 = \log C_1 + b \log 100$$

Solving the equations simultaneously yields:

$$b = \frac{\log 20 - \log 15}{\log 50 - \log 100} = -0.415$$

Thus,

$$p = (2)^{-0.415} = 0.75$$

That is a 75% learning rate. In general, the learning curve exponent, b, may be calculated directly from actual data or computed analytically. That is:

$$b = \frac{\log A_{x_1} - \log A_{x_2}}{\log x_1 - \log x_2}$$

or

$$b = \frac{\ln(p)}{\ln(2)}$$

where:

x_1 = first production level
x_2 = second production level
A_{x_1} = cumulative average cost per unit at the first production level
A_{x_2} = cumulative average cost per unit at the second production level
p = learning rate percentage

Using the basic cumulative average cost function, the total cost of producing x units is computed as:

$$TC_x = (x)A_x = (x)C_1 x^b = C_1 x^{(b+1)}$$

The unit cost of producing the xth unit is given by:

$$U_x = C_1 x^{(b+1)} - C_1 (x-1)^{(b+1)}$$

$$= C_1 \left[x^{(b+1)} - (x-1)^{(b+1)} \right]$$

The marginal cost of producing the xth unit is given by:

$$MC_x = \frac{d[TC_x]}{dx} = (b+1)C_1 x^b$$

Computational example

Suppose in a production run of a certain product it is observed that the cumulative hours required to produce 100 units is 100,000 hours with a learning curve effect of 85%. For project planning purposes, an analyst needs to calculate the number of hours spent in producing the fiftieth unit. Following the notation used previously, we have the following information:

$$p = 0.85$$

$$X = 100 \text{ units}$$

$$Ax = 100,000 \text{ hours} / 100 \text{ units} = 1,000 \text{ hours} / \text{ unit}$$

Now,

$$0.85 = 2^b$$

Therefore, $b = 0.2345$. Also,

$$1000 = C_1 (100)^b$$

Therefore, $C1 = 2{,}944.42$ hours. Thus,

$$C_{50} = C_1 (50)^b$$

$$= 1{,}176.50 \text{ hours.}$$

That is, the cumulative average hours for 50 units is 1,176.50 hours. Therefore, cumulative total hours for 50 units $= 58{,}824.91$ hours. Similarly,

$$C_{49} = C_1 (49)^b = 1{,}182.09 \text{ hours.}$$

That is, the cumulative average hours for 49 units is 1,182.09 hours. Therefore, cumulative total hours for 49 units $= 57{,}922.17$ hours. Consequently, the number of hours for the fiftieth unit is given by:

$$C_{50} - C_{49} = 58{,}824.91 \text{ hours} - 57{,}922.17 \text{ hours}$$

$$= 902.74 \text{ hours.}$$

The unit cost model

The unit cost model is expressed in terms of the specific cost of producing the xth unit. The unit cost formula specifies that the individual cost per unit will decrease by a constant percentage as cumulative production doubles. The formulation of the unit cost model is presented below. Define the average cost as A_x.

$$A_x = C_1 x^b$$

The total cost is defined as:

$$TC_x = (x)A_x = (x)C_1 x^b = C_1 x^{(b+1)}$$

and the marginal cost is given by:

$$MC_x = \frac{d[TC_x]}{dx} = (b+1)C_1 x^b$$

This is the cost of one specific unit. Therefore, define the unit cost model as

$$U_x = (1+b)C_1 x^b$$

where U_x is the cost of producing the xth unit. To derive the relationship between A_x and U_x:

$$U_x = (1+b)C_1 x^b$$

$$\frac{U_x}{(1+b)} = C_1 x^b = A_x$$

$$A_x = \frac{U_x}{(1+b)}$$

$$U_x = (1+b)A_x$$

To derive an expression for finding the cost of the first unit, C_1, we will proceed as follows. Since $A_x = C_1 x^b$, we have:

$$C_1 x^b = \frac{U_x}{(1+b)}$$

$$\therefore C_1 = \frac{U_x x^{-b}}{(1+b)}$$

For the case of continuous product volume (e.g., chemical processes), we have the following corresponding expressions:

$$TC_x = \int_0^x U(z)\,dz = C_1 \int_0^x z^b\,dz = \frac{C_1 x^{(b+1)}}{b+1}$$

$$Y_x = \left(\frac{1}{x}\right)\frac{C_1 x^{(b+1)}}{b+1}$$

$$MC_x = \frac{d[TC_x]}{dx} = \frac{d\left[\dfrac{C_1 x^{(b+1)}}{b+1}\right]}{dx} = C_1 x^b$$

Productivity calculations using learning curves

Let

a_1 = time to complete the task the first time
a_n = overall average unit time after completing the task times
I = time improvement factor
k = learning factor
n = number of times the task has been completed

r_n = ratio of time to perform the task for the *n*th time divided by time to perform for the (*n-1*)th time

t_n = time to perform the task the *n*th time

t_{tn} = total time to perform the task *n* times

Determining average unit time:

$$a_n = \frac{a_1}{n^k}$$

Calculation of the learning factor:

$$k = \frac{\log a_1 - \log a_n}{\log n}$$

Calculating total time for a job:

$$t_{tn} = t_1 n^{1-k}$$

Time required to perform a task for the *n*th time:

$$t_n = a_1 \left(\frac{1-k}{n^k} \right)$$

The improvement ratio

This is the number of times a task must be completed before improvement flattens to a given ratio.

$$n \geq \frac{1}{1 - r^{1/k}}$$

Computational example

Suppose a new employee can perform a task for the first time in 30 minutes. Assume that the learning factor for the task is 0.07. At steady-state operation, an experienced employee can perform the task such that there is less than 1% improvement in successive executions of the task. This means that the ratio of times between succeeding tasks will be 0.99 or larger. How long will it take the experienced employee to perform the task?

Solution

First, determine how many times the task must be performed before we can achieve an improvement ratio of 0.99. Then, use the resulting value of *n* to compute the required processing.

$$n \ge \frac{1}{1 - 0.99^{1/.07}}$$

$$\ge 7.4769 \approx 8$$

Now, determine how long it should take to perform the task for the 8th time.

$$t_8 = 30\left(\frac{1 - 0.07}{8^{.07}}\right)$$

$$= 24.12 \text{ minutes}$$

Computation for improvement goal in unit time

This to find how many times a task should be performed before there will be a given improvement in unit time.

$$n_2 = \frac{n_1}{I^{1/k}}$$

If this computation results in an extremely large value of n_2, it implies that the specified improvement goal cannot be achieved. This could be because the limit of improvement is reached due to leveling off of the learning curve before the goal is reached.

Computation of machine output capacities

Notations

e_c = efficiency of capital equipment in percent of running time
f_a = fraction of output accepted = U_a / U
f_r = fraction of output rejected = U_r / U
N = number of machines
t = time to manufacture one unit (in minutes)
T_a = time per shift that machine actually produces (in minutes)
T_t = hours worked per day = eight times numbers of shifts (i.e., 8 hours per shift)
T_y = production hours per year, usually 2,080 hours (i.e., 8 hrs/day × 260 days)
U = unit per shift = $U_a + U_r$
U_a = units accepted per shift
U_r = units rejected per shift
U_y = units manufactured per year

Machine utilization ratio: Determining how often a machine is idle

$$e_c = \frac{\text{time the machine is working}}{\text{time the machine could be working}}$$

The above expression can also be used for calculating the portion of time a machine is idle. In that case, the numerator would be idle time instead of working time.

Calculation for number of machines needed to meet output

When the total output is specified, the number of machines required can be calculated from this formula:

$$n = \frac{1.67tU}{T_t e_c}$$

Example

Suppose we desire to produce 1,500 units per shift on machines that each complete one unit every five minutes. The production department works one 8-hour shift with two 15-minute breaks. Thus, the machines run seven hours and thirty minutes per day. How many machines will be required?

Solution

Efficiency must first be calculated as a percentage of operating time in relation to shift duration. That is,

$$e_c = \left(\frac{7\,\text{hours, 30\,minutes}}{8\,\text{hours}} \right) 100$$

$$= 93.75\,\text{percent}$$

The above percent efficiency value is now substituted into the computational formula for the number of machines. That is,

$$n = \frac{1.67(5)1,500}{8(93.75)}$$

$$= 16.7 \approx 17\,\text{machines}$$

Alternate forms for machine formula

The formula for calculating number of machines can be rearranged to produce the alternate forms presented below. In some cases, e_c

(percentage efficiency of time usage) is replaced by actual running time in minutes (t_a).

$$t = \frac{0.6\,n\,T_t e_c}{U}$$

$$U = \frac{0.6\,n\,T_t e_c}{t}$$

$$T_t = \frac{1.67\,t U}{n\,e_c}$$

$$e_c = \frac{1.67\,t U}{n\,T_t}$$

$$t = \frac{60\,n\,t_a}{U}$$

$$U = \frac{60\,n\,t_a}{t}$$

$$T_a = \frac{0.0167\,t U}{n}$$

Example

How many hours a day must you operate six machines that each produce one unit every 50 seconds if you need 4,000 units per day? To find hours per day (T_a), substitute directly into the last one of the alternate forms of the machine equation. Note that t is defined in minutes in the equation, but the problem gives the time in seconds. So, first, convert 50 seconds to minutes. That is, 50 seconds = 50/60 minutes. Now, hours per day is calculated as:

$$T_a = \frac{0.0167\,(50/60)\,4,000}{6}$$

$$= 9.28\,\text{hours}$$

Based on this calculation, it is seen that adding more machines is not an option. We must either extend the shift to about 9.28 hours or perform part of the production on another shift, or outsource.

Referring back to the first computational example, let us suppose we want to compare total production from 17 and 18 machines. The units produced per shift can be calculated as shown below:

$$U = \frac{0.6\,(17)\,8\,(93.75)}{5}$$

$$= 1,530\,\text{units from 17 machines}$$

$$U = \frac{0.6(18)8(93.75)}{5}$$

$$= 1,620 \text{ units from } 18 \text{ machines}$$

Calculating buffer production to allow for defects

If a certain fraction of output is usually rejected, a buffer can be produced to allow for rejects. The following formula shows how many to produce in order to have a given number of acceptable units.

$$U = \frac{U_a}{1 - f_r}$$

$$= \frac{U_a}{f_a}$$

Example

A production department normally rejects 3% of a machine's output. How many pieces should be produced in order to have 1,275 acceptable ones? The reject rate of 3% translates to the fraction 0.03. So, $f_r = 0.03$. Thus, we have:

$$U = \frac{1,275}{1 - .03}$$

$$= 1,315$$

Adjusting machine requirements to allow for rejects

To determine how many machines are required for a given amount of shippable output, considering that a given portion of the product is rejected, we will use the following formula:

$$n = \frac{U_y t}{0.6e_c T_y (1 - f_r)}$$

Example

A manufacturer of widgets must have 500,000 ready to ship each year. One out of 300 generally has a defect and must be scrapped. The production machines, each of which makes 50 widgets per hour, work 90% of the time and the plant operates on a 2,080-hour year. Calculate the number of machines required to meet the target production.

Solution

Production time per widget is 60/50 minutes per widget.

$$n = \frac{500,000(60/50)}{0.6(90)2080(1-1/300)}$$

$$= 4.47 \approx 5 \text{ machines}$$

The plant will need five machines, but that will change the utilization ratio. To calculate the new ratio, we will rearrange the formula to solve for *e*.

$$e = \frac{U_y t}{0.6 n T_y (1 - f_r)}$$

$$= \frac{500,000(60/50)}{0.6(5)2080(1-1/300)}$$

$$= 96.48\%$$

The higher percentage utilization is due to using the whole number of machines (5) instead of the fractional computation of 4.47.

Output computations for sequence of machines

In a sequence of machines, the output of one set of machines will move on to the next stages for other sets of machines. The goal here is to determine how many items should be started at the first set of machines so that a given number of units will be produced from the sequence of machines. For each state,

$$\text{Number of units needed at first stage} = \frac{\text{Output needed}}{1 - f_r}$$

Example

The first operation grinds a casting, and 1% of the machined items are rejected instead of being passed on to the next operation. Buffing and polishing are second; 0.5% of the output from this set of machines is rejected. Problems with the third set of machines (nameplate attachment) result in a rejection rate of 2%. How many units should be started into the sequence if we are to ship 30,000 and if all rejects are scrapped?

Apply the formula to each stage, starting at the last one.

$$Start_3 = \frac{30,000}{1-0.02}$$

$$= 30,613$$

Then determine how many units should be started at the second stage in order to have 30,613 units at the input to the third stage.

$$Start_2 = \frac{30,613}{1-.005}$$

$$= 30,767$$

Finally, determine how many units should be started at the first stage.

$$Start_1 = \frac{30,767}{1-0.01}$$

$$= 31,078$$

Therefore, at least 31,078 units should be started at the first stage of this manufacturing process.

Calculating forces on cutting tools

This section gives formulas for finding forces required by milling machines, drills, and other tools that cut or cut into the material (Moffat, 1987). Machine setting for minimum cost is also included. The following notations are used in this section.

a = cross-sectional area of chip or piece, square inches = depth of cut times feed (one spindle revolution)
c_l = cost of rejecting piece outside lower limit
c_u = cost of rejecting piece outside upper limit
d = diameter of work or tool
g = goal for machine setting, inches
k = constant, determined from tables by the formulas
L_l = lower limit for accepting piece
L_u = upper limit for accepting piece
p = pressure at point of cut, pounds
s = cutting speed, feet per minute
T = tolerance, inches
w = horsepower at tool
σ = standard deviation

Calculating pressure

A cutting tool is subjected to pressure at the point of cut, as calculated from the following:

$$p = 80,300(1.33)^k a$$

Where k is a constant, provided in Table 5.1.

Example

A cylinder of cast iron is being cut by a lathe to a depth of 0.06 inch, with a feed of 0.0286 inch per revolution. How much pressure is at the cut?

$$p = 80,300(1.33)^{1.69}(.06)(.0286)$$

$$= 223 \text{ pounds}$$

Finding required horsepower

A formula similar to the formula for pressure finds the horsepower required at the tool's cutting edge.

$$w = 2,433(1.33)^k (a)(s)$$

where k is a constant given in Table 5.1.

Example

A tool cuts mild steel to a depth of 0.045 inches, feeding at 0.020 inches. What horsepower is required at the cutting edge if the cutting speed is 70 feet per minute?

Table 5.1 Constant for pressure and horsepower formulas

Type of material	Constant k
Bronze	1.26
Cast iron	1.69
Cast steel	2.93
Mild Steel	4.12
High carbon steel	5.06

$$w = 2.433(1.33)^{4.12}(.045)(.020)(70)$$

$$= 0.4963 \text{ horsepower}$$

Machine accuracy

The dimensional tolerance to which a piece can be held is a function of both the machining operation and the piece's dimension.

$$T = kd^{0.371}$$

Coefficient k is a function of the machining operation, as shown in Table 5.2.

Example

A bar with 0.875-inch diameter is being turned on a turret lathe. What is a reasonable tolerance?

$$T = .003182(0.875)^{0.371}$$

$$= .003028 \text{ inch}$$

Calculating the goal dimension

The formula below finds the nominal measurement for machining. This may not be the midpoint of the tolerance band. This setting will minimize costs under the following conditions:

1. If the part has not been machined enough, it can be reworked.
2. If the part has been machined too far, it must be scrapped.

   Work coming from an automatic machine can be expected to have a normal distribution about the center value to which the machine is set.

Table 5.2 Constant k for machining accuracy formula

Machining operation	k
Drilling, rough turning	.007816
Finish turning, milling	.004812
Turning on turret lathe	.003182
Automatic turning	.002417
Broaching	.001667
Reaming	.001642
Precision turning	.001378
Machine grinding	.001008
Honing	.000684

However, it is not always desirable for the distribution to be centered about the design value because the cost of correcting rejects may not be the same on each side. This formula applies when it costs more to reject a piece because of too-small dimensions than because of too-large dimensions. It will therefore tell us to set the machine for a goal dimension on the large side of center, so that most of the rejects will be due to too-large dimensions (Moffat, 1987).

$$g = \frac{1}{2(L_u - L_l)} \left[L_u^2 - L_l^2 + \sigma^2 \ln\left(\frac{c_l}{c_u}\right) \right]$$

Example

A solid cylinder is to be turned down to 1.125 inches diameter by a machine under robotic control. If a piece comes off the machine measuring 1.128 or larger, a new shop order must be cut to have the piece machined further. That procedure costs an additional $10.80. When a piece measures 1.122 or smaller, it must be scrapped at a total cost of $23.35. Standard deviation of the machined measurements is 0.002 inch. To what dimension should the machine be set?

$$g = \frac{1}{2(1.128 - 1.122)} \left[1.128^2 - 1.122^2 + .002^2 \ln\frac{23.35}{10.80} \right]$$

$$= 1.1253 \text{ inches}$$

Because it costs more to scrap a piece than to rework it, the calculation above suggests aiming on the high side to avoid undercutting a piece. This is a reasonable practice.

The following notations are used in the next section.

d = diameter of drill, inches
f = feed per revolution, inches
f_{hp} = horsepower equivalent of torque
f_{th} = thrust, pounds
f_{tq} = torque, inch-pounds
v = rotational velocity, rotations per minute (rpm)

Drill thrust

The amount of thrust required for drilling is found from the following formula.

$$f_{th} = 57.5 f^{0.8} d^{1.8} + 625 d^2$$

Example

A 3/8-inch drill is feeding at 0.004 inches per revolution. What is the thrust?

$$f_{th} = 57.5(.004)^{0.8}(0.375)^{1.8} + 625(0.375)^{2}$$

$$= 88.01 \text{ pounds}$$

Drill torque

This formula calculates the torque for a drilling operation.

$$f_{tq} = 25.2f^{0.8}d^{1.8}$$

Example

Calculate the torque for the preceding example.

$$f_{tq} = 25.2(.004)^{0.8}(0.375)^{1.8}$$

$$= 0.052 \text{ inch-pound}$$

Drill horsepower

This formula finds the horsepower equivalent of a torque.

$$f_{hp} = 15.87 \times 10^{-6} f_{tq} v$$

$$= 400 \times 10^{-6} f^{.8}d^{1.8}v$$

Example

What is the equivalent horsepower in the above example if the drill rotates at 650 rpm?

$$f_{hp} = 15.87 \times 10^{-6}(0.052)650$$

$$= 0.0005 \text{ horsepower}$$

Calculating speed and rotation of machines

This section looks at turning speed, surface speed, and rotational speed of tools and work. The notations used here are:

a_l = linear feed, or advance, in inches per minute
a_r = feed, in inches per revolution

a_t = feed, in inches per tooth
d = diameter of work where tool is operating, inches
r = radius of work where tool is operating, inches
s = cutting speed, feet per minute
s_r = rotational speed of work, revolutions per minute
s_s = surface speed of work, feet per minute

Shaft speed

Cylindrical shapes passing a contact point with a tool, as in a lathe, should move at a speed recommended by the manufacturer. This formula translates surface speed of the cylinder to revolutions per minute.

$$s_r = \frac{12s_s}{\pi d}$$

$$= \frac{6s_s}{\pi r}$$

Example

A solid metal cylinder with 1–1/4-inch diameter is to be worked in a lathe. Its recommended surface speed is 240 feet per minute. How fast should the lathe turn?

$$s_r = \frac{12(240)}{3.1416(1.25)}$$

$$= 733 \text{ revolutions per minute}$$

Note that the nearest available speed to 733 rpm will bring the surface speed to its recommended value.

Example

A lathe cuts at a speed of 102 feet per minute on cylindrical work with a diameter of 4.385 inches. What is the spindle speed?

$$s_r = \frac{12(102)}{3.1416(4.385)}$$

$$= 88.85 \text{ revolutions per minute}$$

Example

The material should not be cut faster than 200 feet per minute. What is the fastest rotation allowed when the material's radius is 2–1/2 inches?

$$S_r = \frac{1.9099(200)}{2.25}$$

$$= 152.79 \text{ revolutions per minute}$$

Thus, rotational speed should be kept below 153 feet per minute.

Surface speed

This formula finds the rate at which the work's surface passes the tool.

$$s_s = 0.2618ds_r$$

$$= 0.5236rs_r$$

Example
Cylindrical material of 1.12-inch diameter rotates at 460 rpm. What is the work's surface speed?

$$s_s = 0.2618(1.12)460$$

$$= 134.88 \text{ feet per minute}$$

Tool feed per revolution

The rate at which the tool advances into or past the work is given by this formula.

$$a_r = \frac{a_l}{s_r}$$

$$= \frac{0.2618\,da_l}{s_s}$$

$$= \frac{0.5236\,ra_l}{s_s}$$

Example
The tool is to advance 40 inches per minute when the work has a diameter of 0.875 inches. If the surface speed is 180 feet per minute, at what rate should the tool advance?

$$a_r = \frac{0.2618(0.875)40}{180}$$

$$= 0.0509 \text{ inch per revolution}$$

Tool feed per minute

The formula below is a rearrangement of an earlier example. This is used when the rotational speed is known.

$$a_l = a_r s_r$$

$$= \frac{3.8197 a_r s_s}{d}$$

$$= \frac{1.9099 a_r s_s}{r}$$

Example

Assume that a tool advances 0.075 inches into a work-piece for each revolution and the work is rotating 660 revolutions per minute. What is the feed rate?

$$a_l = 0.075(660)$$

$$= 49.5 \text{ inches per minute}$$

Tool feed per tooth

This formula is used when the cutter information is known in terms of teeth instead of time.

$$a_t = \frac{a_r}{t}$$

$$= \frac{a_l}{t s_r}$$

Example

A cutter with 33 teeth advances 0.06 inch per revolution. At what rate does it advance?

$$a_t = \frac{0.06}{33}$$

$$= 0.0018 \text{ inch per tooth}$$

Computation to find the volume of material removed

Formulas in this section (Moffat, 1987) find the volume of material removed by operations such as cutting, drilling, and milling. In many applications, the results of these formulas (volume) will be multiplied by unit weight of the material to find the weight of material removed. The notations used are:

d = diameter of hole or tool, inches
e = distance cut extends on surface, inches
f = linear feed rate, inches per minute
h = depth of cut, inches
t = angle of tip of drill, degrees
v = volume, cubic inches
v_r = rate of volume removal, cubic inches per minute
w = width of cut, inches

Time rate for material removal in a rectangular cut

This formula gives the rate at which material is removed when the cut has a rectangular shape.

$$v_r = whf$$

Example

A milling machine is cutting a 1/8-inch deep groove with the face of a 5/8-inch diameter tool. It feeds at 28 inches per minute. At what rate is material removed?

$$v_r = 0.625(0.125)28$$

$$= 2.1875 \text{ cubic inches per minute}$$

Calculation of linear feed for rectangular material removal

This formula is used to find the feed rate required for a given rate of material removal.

$$f = \frac{v_r}{wh}$$

This formula applies when material is removed by a drill. One version of the formula approximates the volume by assuming the drill is a plain cylinder without a triangular tip. The second version is precise and includes a correction for the angle of the drill's tip.

Example

Let's assume that we want to remove three cubic inches per minute while milling 1/8-inch from the side of a block that is 1–1/16 inches thick. How fast should the cutter feed if we are cutting with the circumference of a tool at least 1–1/16 inches long?

$$f = \frac{3}{1.0625(0.125)}$$

$$= 22.6 \text{ inches per minute}$$

Takt time for production planning

"Takt" is the German word referring to how an orchestra conductor regulates the speed, beat, or timing so that the orchestra plays in unison. So, the idea of takt time is to regulate the rate time or pace of producing a completed product. This refers to the production pace at which workstations must operate in order to meet a target production output rate. The production output rate is set based on product demand. In a simple sense, if 2,000 units of a widget are to be produced within an 8-hour shift to meet a market demand, then 250 units must be produced per hour. That means, a unit must be produced every 60/250 = 0.24 minutes (14.8 seconds). Thus, the **takt time** is 14.–8 seconds. Lean production planning then requires that workstations be balanced such that the production line generates a product every 14.4 seconds. This is distinguished from the **cycle time**, which refers to the actual time required to accomplish each workstation task. Cycle time may be less than, more than, or equal to takt time. Takt is not a number that can be measured with a stop watch. It must be calculated based on the prevailing production needs and scenario. Takt time equation is:

$$Takt\ time = \frac{(Available\ work\ time) - (Breaks)}{Customer\ Demand}$$

$$= \frac{Net\ Available\ Time\ per\ Day}{Customer\ Demand\ per\ Day}$$

Takt time is expressed as "seconds per piece," indicating that customers are buying a product once every so many seconds. Takt time is not expressed as "pieces per second."

The objective of lean production is to bring the cycle time as close to the takt time as possible; that is choreographed. In a balanced line design, the takt time is the reciprocal of the production rate.

Improper recognition of the role of takt time can cause an analyst to overestimate the production rate capability of a line. Many manufacturers have been known to over-commit to customer deliveries without accounting for the limitations imposed by takt time. Since takt time is set based on customer demand, its setting may lead to an unrealistic expectation of workstations. For example, if the constraints of the prevailing learning curve will not allow sufficient learning time for new operators, then takt times cannot be sustained. This may lead to the need for buffers to temporarily accumulate units at some workstations. But this defeats the pursuits of lean production or just-in-time. The need for buffers is a symptom of imbalances in takt time. Some manufacturers build **takt gap** into their production planning for the purpose of absorbing non-standard occurrences in the production line. However, if there are more non-standard or random events than have been planned for, then production rate disruption will occur.

It is important to recognize that the maximum production rate determines the minimum takt time for a production line. When demand increases, takt time should be decreased. When demand decreases, takt time should be increased. Production crew size plays an important role in setting and meeting takt time. The equation for calculating the crew size for an assembly line doing one-piece flow that is paced to takt time is presented below:

$$Crew\ size = \frac{Sum\ of\ Manual\ Cycle\ Time}{Takt\ Time}$$

Production crew work-rate analysis

When resources work concurrently at different work rates, the amount of work accomplished by each may be computed for work planning purposes. The general relationship between work, work rate and time can be expressed as:

$$w = rt$$

where:

w = Amount of actual work accomplished. This is expressed in appropriate units, such as miles of road completed, lines of computer code entered, gallons of oil spill cleaned, units of widgets produced or surface area painted.

r = Work rate per unit time at which the assigned work is accomplished.

t = Total time required to accomplish the work.

It should be noted that work rate can change due to the effects of learning curves. In the discussions that follow, it is assumed that work rates remain constant for at least the duration of the work being analyzed.

Work is defined as a physical measure of accomplishment with uniform density (i.e., homogeneous). For example, one square footage of construction may be said to be homogeneous if one square footage is as complex and desirable as any other square footage. Similarly, cleaning one gallon of oil spill is as good as cleaning any other gallon of oil spill within the same work environment. The production of one unit of a product is identical to the production of any other unit of the product. If uniform work density can be assumed for the particular work being analyzed, then the relationship is defined as one whole unit, and the tabulated relationship in Table 5.3 will be applicable for the case of a single resource performing the work.

Where $1/x$ is the amount of work accomplished per unit time. For a single resource to perform the whole unit of work, we must have the following:

$$\left(\frac{1}{x}\right)(t) = 1.0$$

That means the magnitude of x must equal the magnitude of t. For example, if a construction worker can build one block in 30 minutes, then his work rate is 1/30 of a block per minute. If the magnitude of x is greater than the magnitude of t, then only a fraction of the required work will be performed. The information about the proportion of work completed is useful for resource planning and productivity measurement purposes.

Production Work Rate Example

A production worker can custom-build 3 units or a product every 4 hours. At that rate how long will it take to build 5 units? From the information given, we can write the proportion 3 units is to 4 hours as 5 units is to x hours, where x represents the number of hours the worker will take to build 5 units. This gives:

$$\frac{3\,units}{4\,hours} = \frac{5\,units}{x\,hours}$$

which simplifies to yield, $x = 6$ hours 40 minutes.

Case of multiple resources working together

In the case of multiple resources performing the work simultaneously, the work relationship is as presented in Table 5.4.

Table 5.3 Work-rate table for single resource unit

Resource	Work Rate	Time	Work Done
Resource Unit	1/x	1	1.0

Table 5.4 Work-rate table for multiple resource units

Resource type i	Work rate, r_i	Time t_i	Work done w_i
RES 1	r_1	t_1	$\dfrac{r_1}{t_1}$
RES 2	r_2	t_2	$\dfrac{r_2}{t_2}$
...	...	...	...
RES n	r_n	t_n	$\dfrac{r_n}{t_n}$
Total			**1.0**

For multiple resources or work crew types, we have the following expression:

$$\sum_{i=1}^{n} r_i t_i = 1.0$$

where:

n = number of different crew types
r_i = work rate of crew type i
t_i = work time of crew type i

The expression indicates that even though the multiple crew types may work at different rates, the sum of the total work they accomplished together must equal the required whole unit (i.e., the total building). For partial completion of work, the expression becomes:

$$\sum_{i=1}^{n} r_i t_i = p$$

where p is the percent completion of the required work.

Computational examples

Suppose RES 1, working alone, can complete a construction job in 50 days. After RES 1 has been working on the job for 10 days, RES 2 was assigned to help RES 1 in completing the job. Both resources working together finished the remaining work in 15 days. It is desired to determine the work rate of RES 2.

The amount of work to be done is 1.0 whole unit. The work rate of RES 1 is 1/50 of construction work per unit time. Therefore, the amount of work completed by RES 1 in the 10 days it worked alone is (1/50)(10)=1/5 of the required work. This may also be expressed in terms of percent completion or earned value using C/SCSC (cost/schedule control systems criteria). The remaining work to be done is 4/5 of the total work. The two resources working together for 15 days yield the analysis shown in Table 5.5.

Thus, we have $15/50 + 15r_2 = 45$, which yields $r_2 = 1/30$ for the work rate of RES 2. This means that RES 2, working alone, could perform the construction job in 30 days. In this example, it is assumed that both resources produce identical quality of work. If quality levels are not identical for multiple resources, then the work rates may be adjusted to account for the different quality levels or a quality factor may be introduced into the analysis.

As another example, suppose the work rate of RES 1 is such that it can perform a certain task in 30 days. It is desired to add RES 2 to the task so that the completion time of the task could be reduced. The work rate of RES 2 is such that it can perform the same task alone in 22 days. If RES 1 has already worked 12 days on the task before RES 2 comes in, find the completion time of the task. It is assumed that RES 1 starts the task at time zero.

As usual, the amount of work to be done is 1.0 whole unit (i.e., the full construction work). The work rate of RES 1 is 1/30 of the task per unit time and the work rate of RES 2 is 1/22 of the task per unit time. The amount of work completed by RES 1 in the 12 days it worked alone is $(1/30)(12) = 2/5$ (or 40%) of the required work. Therefore, the remaining work to be done is 2/5 (or 60%) of the full task. Let T be the time for which both resources work together. The two resources, working together, to complete the task yield the results shown in Table 5.6.

Thus, we have $T/30 + T/22 = 3/5$ which yields $T = 7.62$ days. Consequently, the completion time of the task is $(12 + T) = 19.62$ days from time zero. It is assumed that both resources produce identical quality of work and that

Table 5.5 Work-rate analysis for construction example

Resource type i	Work rate, r_i	Time t_i	Work done w_i
RES 1	**1/50**	**15**	**15/50**
RES 2	r_2	**15**	$15r_2$
Total			**4/5**

Table 5.6 Work-rate table for alternate work example

Resource type i	Work rate r_i	Time t_i	Work done w_i
RES 1	**1/30**	**T**	**T/30**
RES 2	**1/22**	**T**	**T/22**
Total			**4/5**

Table 5.7 Incorporation of cost into work-rate table

Resource (i)	Work rate r_i	Time t_i	Work done w_i	Pay rate p_i	Total cost C_i
Crew A	r_1	t_1	$\dfrac{r_1}{t_1}$	p_1	C_1
Crew B	r_2	r_2	$\dfrac{r_2}{t_2}$	p_2	C_2
...	...	...	...	...	...
Crew n	r_n	r_n	$\dfrac{r_n}{t_n}$	p_n	C_n
Total			1.0		Total Cost

the respective work rates remain consistent. The relative costs of the different resource types needed to perform the required work may be incorporated into the analysis as shown in Table 5.7.

Calculation of personnel requirements

The following expression calculates personnel requirements in a production environment.

$$N = \sum_{i=1}^{n} \frac{T_i O_i}{eH}$$

where:
   N = number of production employees required
   n = number of types of operations
   T_i = standard time required for an average operation in O_i
   O_i = aggregate number of operation type i required on all the products produced per day
   e = assumed production efficiency of the production facility
   H = total production time available per day

Calculation of machine requirements

The following expression calculates machine requirements for a production operation.

$$N = \frac{t_r P}{t_a e}$$

where:
   N = number of machines required
   t_r = time required in hours to process one unit of product at the machine

t_a = time in hours for which machine is available per day
P = desired production rate in units per day
e = efficiency of the machine

References

Badiru, A. B. (2012). *Project Management: Systems, Principles, and Applications*, Taylor & Francis Group/CRC Press, Boca Raton, FL.

Badiru, A. B., S. Abidemi Badiru, and I. Adetokunbo (2008). *Industrial Project Management: Concepts, Tools, and Techniques*, Taylor & Francis Group/CRC Press, Boca Raton, FL.

Badiru, A. B. and O. A. Omitaomu (2007). *Computational Economic Analysis for Engineering and Industry*, Taylor & Francis Group/CRC Press, Boca Raton, FL.

Badiru, A. B. and O. A. Omitaomu (2011). *Handbook of Industrial Engineering Equations, Formulas, and Calculations*, Taylor & Francis Group/CRC Press, Boca Raton, FL.

Heragu, S. (1997). *Facilities Design*, PWS Publishing Company, Boston, MA.

Moffat, D. W. (1987). *Handbook of Manufacturing and Production Management Formulas, Charts, and Tables*, Prentice-Hall, Englewood Cliffs, NJ.

chapter six

Introduction to project management

Introduction

There is a growing need for better project management, especially for information technology (IT) projects. Project management is regarded as high priority as all companies or organizations, whether small or large, are at one time or the other involved in implementing undertakings, innovations, and changes. Project management ensures the project keeps to task, that potential issues are quickly eliminated and projects are delivered on time, all the while making sure everyone knows what is happening and the project quality and budget are acceptable.

Civilization has given rise to huge corporations and projects of unsurpassed size and scope. Traditional project management often focuses on small projects, which can be completed in a short time period. However, larger projects and projects that are part of larger programs, such as those in the oil and gas industry, requires enterprise project management (EPM), and there is the need to know what it means to a project team.

Large companies and organizations may work on dozens of projects at the same time, and these projects may be spread over multiple departments or locations. This collection of projects is called the company project portfolio. Since projects may interact with each other or be affected by other company operations, managing this portfolio can become a complex task. EPM effectively manages a company project portfolio. There is a balancing of interdependent elements and details that are important to the business or industry in which they operate. The market environment surrounding each project and how each project supports the company mission or goal is known. EPM can identify when projects conflict or duplicate efforts, and they can recognize underfunded projects that might present an opportunity for the organization. Companies not employing EPM may not be able to act quickly enough to stop loss or capitalize on a potential window of opportunity.

The US spends $2.3 trillion on projects every year, an amount equal to one-quarter of the nation's gross domestic product (GDP), the world as a whole spends nearly $10 trillion of its $40.7 trillion gross product on projects of all kinds, more than 16 million people regard project management as their profession; on average, a project manager earns more than $82,000 per year. More than half a million new IT application development projects were initiated in the US during 2001, up from 300,000 in 2000.

What is a project?

A project is a temporary endeavor undertaken to accomplish a unique product or service. A project:

- has a start and end date
- is unique
- brings about change
- has unknown elements, which therefore create risk

The attributes of a project are as follows:

- unique purpose
- temporary
- require resources, often from various areas
- should have a primary sponsor and/or customer
- involve uncertainty

Project stakeholders

Stakeholders are the people involved in or affected by project activities. It refers to an individual, group, or organization, who may affect, be affected by, or perceive itself to be affected by a decision, activity, or outcome of a project. Stakeholders include:

- **Project sponsor**
  Project sponsor is the individual (often a manager or executive) with overall accountability for the project. The project sponsor is primarily concerned with ensuring that the project delivers the agreed business benefits.

- **Project team**
  A project team is a team whose members usually belong to different groups and functions, and are assigned to activities for the same project. A team can be divided into sub-teams according to need. Usually project teams are only used for a defined period of time.

- **Support staff**
  Project support officers provide vital assistance to project managers. These are people or members of the project team who do project management activities like communications, scheduling, and risk management. They are also involved in making the budget and

reports, as well as other types of activities related to administrative support.

- **Customers**
  These are the indirect beneficiaries of the project who are external or outside the company. The project management activities are driven by customers such that they are involved during the start (gathering of requirements) until the very end (delivering products and services). This is the reason why project managers always include customer satisfaction during planning. By keeping customers satisfied through quality product, good communication, and more, businesses can grow.

- **Users**
  The user represents those groups who will use or gain/benefit from the project and must be empowered to make decisions on their behalf. In practice the user is likely to be responsible for realizing the business benefits and may have service commitments after the project is completed.

- **Suppliers**
  The supplier represents those groups who will design, develop, facilitate, procure, and implement the project and must be empowered to make decisions on their behalf. In larger or more complex projects a Supplier Advisory Board can be set up to represent wider supplier interests.

- **Opponents to the project**
  A project manager needs to identify all stakeholders up front and focus not only on those who will support the project, but to think as broadly and deeply as possible about all of those who might stand in its way. In identifying opponents to projects, the following need to be asked:
  - Who might be opposed to my project?
  - What is their relationship(s) with us?
  - What are their spheres, and levels of power and influence?
  - What are their relationships with each other?
  - How might opponents work together in unexpected ways to impede progress?

General overview of project management

Project management is the application of knowledge, skills, tools, and techniques to project activities in order to meet project requirements. Project management can also be defined as the practice of initiating, planning, executing, controlling, and closing the work of a team to achieve specific goals and meet specific success criteria at the specified time. A project is a temporary endeavor designed to produce a unique product, service, or result with a defined beginning and end (usually time-constrained, and often constrained by funding or staffing) undertaken to meet unique goals and objectives, typically to bring about beneficial change or added value.

The aim of project management is to achieve all of the project goals within the given constraints. This information is usually described in project documentation, created at the beginning of the development process. The primary constraints are scope, time, quality, and budget. The secondary, and more ambitious, challenge is to optimize the allocation of necessary inputs and apply them to meet pre-defined objectives. The object of project management is to produce a complete project which complies with the client's objectives. In many cases the object of project management is also to shape or reform the client's brief in order to feasibly be able to address the client's objectives. Once the client's objectives are clearly established they should impact all decisions made by other people involved in the project – project managers, designers, contractors, subcontractors, etc. If the project management objectives are ill-defined or too tightly prescribed it will have a detrimental effect on decision-making. As a discipline, project management developed from several fields of application including civil construction, engineering, and heavy defense activity.

History of project management

Although there has been some form of project management since early civilization, project management in the modern sense began in the 1950s. The 1950s marked the beginning of the modern project management era where core engineering fields came together to work as one. Project management became recognized as a distinct discipline arising from the management discipline with the engineering model. Most people consider the Manhattan Project to be the first project to use "modern" project management. This 3-year, $2 billion (in 1946 dollars) project had a separate project manager and a technical manager. Some people argued that building the Egyptian pyramids was a project, as well as building the Great Wall of China. This time, not only did the Chinese emperor order millions of people to complete the project, but he also made sure that his workforce was divided into three separate groups: soldiers, civilians, and criminals (we guess the latter group got a raw deal). As a discipline, project management developed from several fields of application including civil construction, engineering, and heavy defense activity.

Although project management as we know it today didn't begin until the middle of the twentieth century, it has been present and steadily evolving for thousands of years. One of the earliest forms of project management dates back to ancient Egypt, more specifically to the construction of King Khufu's Great Pyramid at Giza, in 2570 BC. According to historians and ancient records, Egyptians appointed managers to oversee the completion of each of the four sides of the pyramid, and they were also responsible for the planning and the execution of the project.

The first and second Industrial Revolutions brought about significant changes in the development of project management. Although the two revolutions are mainly associated with improved working and living conditions, with urbanization and great engineering works (like the first railway and the completion of the first transcontinental railroad) their effect on project management shouldn't be overlooked. During the Industrial Revolution, industry expanded rapidly all over the world. The beginning of automation and the growth of factories meant that everything was done on a much greater scale. This, in turn, meant that people were able to manage projects in a completely different and more extensive way. The twentieth century saw significant changes in the world of project management. Both Frederick Taylor (often referred to as the father of scientific management) and his friend Henry Gantt played an important role in the way projects were managed – and are still managed today.

Henry Gantt, known as one of the forefathers of project management, is best known for creating and designing his famous diagram, the Gantt chart; a radical idea at the time and an innovation that changed the way projects were managed and documented in the twentieth century. Developed in 1917, the aim of Gantt's chart was to track the progress of ship-building projects during World War I. By documenting and examining each step of the process, he was able to get a clear overall view of the entire project and gather information about the connection between various functions.

The 1950s marked the beginning of the modern project management era, when project management became recognized as a distinct discipline and companies began to apply formal project management tools and techniques to complex projects. One of the most important inventions of the decade was the Critical Path Method (CPM), developed by the DuPont Corporation and Remington Rand Corporation in 1957. The aim of CPM was to assess and calculate the activities required to complete a project and predict the length of each of these phases. The idea was so successful that it is reported to have saved the company $1 million in its first year of implementation. Another significant development from this era is the Program Evaluation and Review Technique (or simply, PERT), developed by the United States Navy as part of the Polaris missile submarine program. Similar to CPM, PERT is used for analyzing the tasks that are required to complete a specific project, as well as estimating the time needed to complete each of these tasks and the project itself. Although the two methods are very similar, there is a critical difference between them. While CPM is used for projects where the time at which each individual task is supposed to be carried out are known, PERT is used for projects where these times are either varied or unknown. Because of this difference, CPM and PERT are used in completely different contexts and are not interchangeable.

Motivation for undertaking IT projects

IT project management is the process of planning, organizing, and delineating responsibility for the completion of an organization's specific IT goals. IT projects have a terrible track record. A 1995 Standish Group study (CHAOS) found that only 16.2% of IT projects were successful and over 31% were canceled before completion, costing over $81 billion in the US alone (The Standish Group, 2001). However, the need for IT projects keeps increasing. In 2000, there were 300,000 new IT projects; in 2001, over 500,000 new IT projects were started.

Losses loom larger than gains, and IT project managers are driven more strongly to avoid losses than to achieve gains. Stakeholders on a project perceive the intended new organizational capability/state as worse/inferior to what exists now. It is the responsibility of the IT project manager to deliver projects on time and in budget, by planning and organizing resources and people. IT project managers need to track work to be completed, set deadlines, and delegate tasks to subordinates.

Examples of IT projects

IT projects refer to projects involving hardware, software, and networks. Organizations develop new software or enhance existing systems to perform many business functions. Many organizations upgrade hardware, software, and networks via projects. The success of an IT project depends on the active engagement and commitment of clients and the project team during the project scope formation, execution, and closing. Clients are asked to:

1. Provide any data, resources, and background information necessary to complete the project
2. Provide a primary contact person(s) that is available for regular status reports, consultation on issues that may arise, and
3. Attend a final presentation of the project report

The following are examples of IT project management projects:

- Northwest Airlines developing a new reservation system called ResNet
- Technology selection and evaluation
- Feasibility studies and proof of concept for new products and services

- Data analytics projects related to demand estimation, diagnostic systems, market
- Competitive intelligence capture and analysis using advanced web crawlers and analytics
- Work-flow analysis and design; process improvement
- Business intelligence/Online analytical processing (OLAP) analysis and design
- Web-based portals and dashboards

Advantages of using formal project management

Formal project management identifies which project(s) is successful, and which need to be halted or modified because they are losing money. Project management is a powerful business tool that can deliver many advantages to businesses of all sizes. It gives businesses repeatable processes, guidelines, and techniques to help manage people and the work involved in projects. It can increase chances of success and help deliver projects consistently, efficiently, on time, and in budget. The main advantage of formal project management is that it helps businesses to manage their projects effectively, enabling them to resolve problems more quickly. It takes time and money to manage a project; however, the following good practices can help businesses:

- Better control of financial, physical, and human resources
- Improved customer relations
- Shorter development times
- Lower costs
- Higher quality and increased reliability
- Higher profit margins
- Improved productivity
- Better internal coordination
- Higher worker morale

The triple constraint

All projects are carried out under certain constraints – traditionally, they are cost, time, and scope. It is the project manager's duty to balance these three often competing goals. These three factors (commonly called "the triple constraint") are represented as a triangle (see Figure 6.1). Projects must:

- be delivered within cost
- be delivered on time

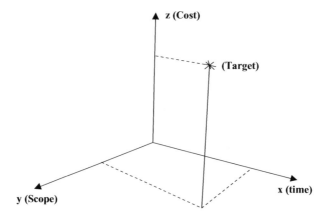

Figure 6.1 The tripe constraint of project management.

- meet the agreed scope – no more, no less
- meet customer quality requirements
  a. **Cost:** What should a project cost? All projects have a finite budget; the customer is willing to spend a certain amount of money for delivery of a new product or service. If a project's cost is reduced, either scope will be reduced or time is increased.
  b. **Time (Schedule):** How long should a project take to complete? Projects have a deadline date for delivery. When a project's time is reduced, either cost is increased or scope is reduced.
  c. **Scope:** What is the project trying to accomplish? Many projects fail on this constraint because the scope of the project is either not fully defined or understood from the start. When a project's scope is increased, either cost or time is increased.
  d. **Quality:** How about standards to meet customers' requirements and quality assurance strategies?

Project management framework

A conceptual framework is provided of how a project can be initiated, planned, executed, and closed within a regular project life cycle. This is shown in Figure 6.2 and each process is described below.

Scope management

Project scope is the part of project planning that involves determining and documenting a list of specific project goals, deliverables, features, functions, tasks, deadlines, and, ultimately, costs. In other words, project scope is what needs to be achieved and the work that must be done to deliver a project.

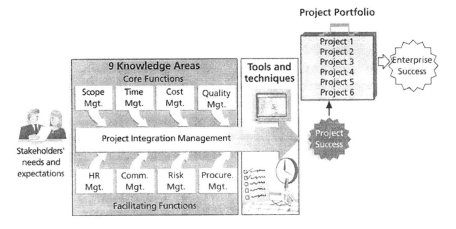

Figure 6.2 Project management framework.

Time management

Time management is the ability to organize and plan the time spent on activities in a day. The result of good time management is increased effectiveness and productivity. It is a key aspect of project management and involves skills such as planning, setting goals, and prioritizing for a better performance.

Cost management

Cost management is concerned with the process of planning and controlling the budget of a project or business. It includes activities such as planning, estimating, budgeting, financing, funding, managing, and controlling costs so that the project can be completed within the approved budget.

Quality management

Quality management is the process for ensuring that all project activities necessary to design, plan, and implement a project are effective and efficient with respect to the purpose of the objective and its performance.

Human resources management

Human resources management is the process of identifying and documenting project roles, responsibilities, required skills, reporting relationships, and creating a staffing management plan.

Communications management

This includes the processes that are required to ensure timely and appropriate planning, collection, creation, distribution, storage, retrieval, management, control, monitoring, and the ultimate disposition of a project.

Risk management

Project risk management is the process of identifying, analyzing, and then responding to any risk that arises over the life cycle of a project to help the project remain on track and meet its goal. A risk is anything that could potentially impact your project's timeline, performance, or budget.

Procurement management

Project procurement management is the creation of relationships with outside vendors and suppliers for goods and services needed to complete a project. This process is comprised of five steps, including initiating and planning, selecting, contract writing, monitoring, and closing and completing.

Improvements in project's success

According to the Standish Group (2001), there has been decided improvement in project success. Time overruns significantly decreased to 163% compared to 222%, cost overruns were down to 145% compared to 189%, required features and functions were up to 67% compared to 61%, 78,000 US projects were successful compared to 28,000, and 28% of IT projects succeeded compared to 16%.

The reasons for the increase in successful projects vary. Some of the reasons for these improvements include:

- The average cost of a project has been more than cut in half.
- Better tools have been created to monitor and control progress
- Better skilled project managers with better management processes are being used.

Project management knowledge areas

Project management knowledge areas describe the key competencies that project managers must develop. Knowledge areas consist of the processes that are applicable to a project as a whole in the most complex instance. It must be noted that, although all projects have to undergo the entire project phase, there are many cases where one or more of the knowledge areas are not applicable. The core knowledge areas lead to specific project objectives, (i.e., scope, time, cost, and quality). The following four facilitating knowledge areas are the means through which the project objectives are achieved:

- **Human resources**, which includes the processes that organize and manage the project team.
- **Communication**, which includes the processes concerning the timely and appropriate generation, collection, dissemination, storage, and ultimate disposition of project information.
- **Risk**, which includes the processes concerned with conducting risk management on a project.
- **Procurement management**, which includes the processes that purchase or acquire products, services or results, as well as contract management processes.

Project management tools and techniques

Project management tools and techniques assist project managers and their teams in various aspects of project management. The tools are classified into the triple constraints, i.e., scope, time, and cost. Some of these tools include:

- **Scope** – Project Charter, Scope Statement, and Work Breakdown Structure (WBS)
- **Time** – Gantt Charts, Network Diagrams, Critical Path Analysis, Critical Chain Scheduling
- **Cost** – Cost estimates and earned value management

Some of the techniques are described below:

Gantt charts and WBS

A Gantt chart is a chart in which a series of horizontal lines shows the amount of work done or production completed in certain periods of time in relation to the amount planned for those periods. The WBS can be defined as a deliverable oriented hierarchical decomposition of the work to be executed by the project team.

As displayed in Figure 6.3, the WBS is shown on the left of the figure. The Gantt chart is shown on the right of the figure. Each task's start and finish date are shown by the Gantt chart using a calendar timescale. Early Gantt charts, first used in 1917, were drawn by hand.

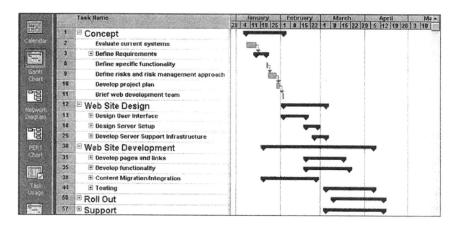

Figure 6.3 Sample Gantt chart in Microsoft Project 2002.

Network diagrams

A network diagram is a graphical way to view tasks, dependencies, and the critical path of a project. Boxes (or nodes) represent tasks and dependencies show up as lines that connect those boxes. A sample network diagram is shown in Figure 6.4. Each box is a project task from the WBS. Arrows show dependencies between tasks. The bolded tasks are on the critical path. If any tasks on the critical path take longer than planned, the whole project will slip unless something is done. Network diagrams were first used in 1958 on the Navy Polaris project, before project management software was available.

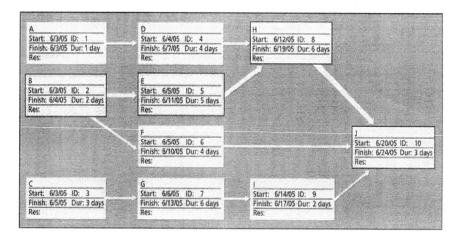

Figure 6.4 Sample Gantt chart in Microsoft Project 2002.

An EPM tool

The EPM tool provides companies with the tools needed to efficiently manage all of a company's projects at an enterprise level. This involves viewing projects from a strategic perspective, allowing executives to prioritize projects and delegate resources, and determining how different projects fit. In recent years, organizations have been taking advantage of software to help manage their projects throughout the enterprise. A sample EPM tool is displayed in Figure 6.5. By 2003, there were hundreds of different products to assist in performing project management. There are three main categories of tools:

- **Low-end tools**: Handle single or smaller projects well.
- **Mid-range tools**: Handle multiple projects and users; Project 2000 is the most popular.
- **High-end tools**: Also called EPM software, often licensed on a per-user basis

Relationship of project management to other disciplines

Project management applies to work as well as personal projects and also to many different disciplines (IT, construction, finance, sports, event

Company ABC Project Portfolio				
Project Name	**Scope**	**Schedule**	**Budget**	**Links**
Active Projects				
Project 1	○	◐	◐	
Project 2	●	●	◐	
Project 3	○		○	
Project 4	○	◐	●	
Approved Projects				
Project 10	○	○	○	
Project 11	○	○	○	
Project 12	○	○	○	
Project 13	○	○	○	
Project 14	○	○	○	
Opportunities				
Project 100				
Project 200				
○	White = going well			
◐	Gray = some problems			
●	Black = major problems			

Figure 6.5 Sample enterprise project management tool.

planning, etc.). Project management skills can help in everyday life. Project management is a relatively new discipline; one that has been formalized only in the last half century or so. Consequently, both academics and practitioners routinely draw upon knowledge in allied (or related) disciplines in order to advance the theory and practice of project management. The knowledge needed to manage projects is unique to the discipline of project management. There are three key disciplines related to project management:

- The business analyst
- The software designer
- The test manager

The business analyst

A business analyst is a person who improves business processes and systems to meet a company's strategic objectives. A business analyst needs good communication and conflict management skills and is always interacting with business users and IT. In smaller projects, a business analyst can sometimes play the role of "assistant project manager" because of their knowledge of the domain and important project/stakeholder issues.

The software designer

A software designer conceptualizes software components at a high-level to make sure they meet end user requirements. A software designer needs skills similar to that of a project manager – communication skills, conflict management skills industry/solution knowledge, and the ability to multi-task.

The test manager

A test manager plans and executes a test plan to ensure the quality of a software product has been verified by its users. Next to the project manager, the test manager is one of the most stressful roles in a project. This is due to the tight timelines for testing and the fact that the test manager is the "control point" before the product gets rolled out.

In addition, project managers must also have knowledge and experience in:

- General management
- Application area of the project

The project management profession is continuously evolving, so the project management community should be receptive to new ideas and also be sensitive to the yearning of the public and professional community so as to model project management practices to meet their expectations.

The project management profession

The job of the IT project manager is in the top ten list of in-demand IT skills (Table 6.1). Professional societies like the Project Management Institute (PMI) have grown tremendously, and project management research and certification programs continue to grow.

Ethics in project management

Just like in project management, ethics is an important part of all professions. Project managers often face ethical dilemmas. Indeed, being ethical and following ethical norms can be said to be among the prerequisites for project managers who have to practice and observe these rules and norms. The importance of ethics for project management can be seen from the way in which project managers have to bid for projects after full disclosure of their capabilities and capacities without resorting to hyperbole or exaggeration during the bidding process; this ensures that they do not employ underhand means to bag the project. Indeed, bidding and prospecting for projects are the primary sources of unethical behavior and unacceptable conduct.

While some might justify the practices such as lobbying, entertaining the clients by wining and dining them, and offering material and non-material inducements to bag the projects as being part of the ways

Table 6.1 Top ten in-demand IT skills

Rank	IT Skill/Job	Average Annual Salary
1	SQL Database Analyst	$80,664
2	Oracle Database Analyst	$87,144
3	C/C++ Programmer	$95,829
4	Visual Basic Programmer	$76,903
5	E-commerce/Java Developer	$89,163
6	Windows NT/2000 Expert	$80,639
7	Windows/Java Developer	$93,785
8	Security Architect	$86,881
9	Project Manager	$95,719
10	Network Engineer	$82,906

Paul Ziv, "The Top 10 IT Skills in Demand," Global Knowledge Webcast *(www.globalknowledge.com)* (11/20/2002).

of doing business, it needs to be remembered that once a project is won based on such methods, the rest of the phases of the project are tainted and compromised as the costs incurred for the above mentioned aspects have to be recovered. This means that the project manager would have to look for ways in which to cut corners as otherwise, the project would become unviable.

Conclusion

Project management requires close scrutiny of current projects to identify their problems and to identify the best solution. An initial implementation planning stage is the first step to identify a solution to the problem. This process will depend on company culture and the ability of the organization to modify and retool the solution to the problem. The solution to the problem should cover the responsible people, indicate major tasks, and list the recommended changes to technology and training. Most companies will set short, medium, and long-term goals.

Acquiring stakeholder involvement is the next step. Each person's role should be clearly identified, including senior management. New information is often discovered at this point that leads to modification of the implementation plan. A list of requirements and phases for implementation is the final step before roll out. The list will include details that all stakeholders have agreed upon (roles, responsibility, and training) and changes to current technology, production, and support.

Implementation of such a comprehensive change can be a significant challenge. A large number of people will have to agree upon changes, and there will be increased project visibility and monitoring. Departments not directly affected by the switch to EPM, such as Human Resources, may need to modify their procedures to support the new model. Roadblocks can be minimized by ensuring clear communication and making sure all changes are realistic.

Reference

The Standish Group. (2001). "CHAOS 2001: A recipe for success."

Bibliography

Cleland, D., and Gareis, R. (2006). Chapter 1: "The evolution of project management" In *Global Project Management Handbook*. Second Edition. McGraw-Hill, New York, 1-3–1-18.

Ferriani, C., Frederiksen, S., and Florian, T. (2011). *Project-Based Organizing and Strategic Management*. Advances in Strategic Management. Emerald Group Publishing, Bingley, UK.

Hornby, R. (2000). Building effective enterprise project management (EPM). Paper presented at *Project Management Institute Annual Seminars & Symposium*, Houston, TX. Newtown Square, PA: Project Management Institute.

Kahneman, D. and Tversky, A. (1979). Prospect theory: An analysis of decision under risk. *Econometrica*, 47(2), 263–291.

Nokes, S. (2007). *The Definitive Guide to Project Management*. Second Edition. Financial Times/Prentice Hall, London.

Phillips, J. (2011). *PMP Project Management Professional Study Guide*. McGraw-Hill Education.

Project Management Institute. (2001). *The PMI Project Management Fact Book*, Second Edition.

Project Management Institute. What Is Project Management? Retrieved 2014-06-04 from https://www.pmi.org/about/learn-about-pmi/what-is-project-management.

chapter seven

Business plan development

Introduction

An important task in starting a new venture is to develop a business plan, which is a "road map" to guide the future of a business or venture. A business plan is a formal statement of business goals, reasons these goals are attainable, and plans for reaching the goals. It may also contain background information about the organization or the project team attempting to reach those goals. Whether starting or growing a business, there is a need for a business plan. The plan will provide the roadmap to achieve business success. The importance of planning should never be overlooked. The elements of the business plan will have an impact on daily decisions and provide direction for expansion, diversification, and future evaluation of the business. For a business to be successful and profitable, the owners and the managing directors must have a clear understanding of the firm's customers, strengths, and competition. They must also have the foresight to plan for future expansion. Whether it's a new business or an existing business in the process of expanding, money is often an issue. Taking time to create an extensive business plan provides an insight into a business.

A business plan is very specific to each particular business. However, while each business needs a unique plan, the basic elements are the same in all business plans. To complete an effective business plan, time must be dedicated to completing the plan. It requires the business to be objective, critical, and focused. The finished project is an operating tool to help manage a business and enable the business to achieve greater success. The plan also serves as an effective communication tool for financing proposals.

An important task in starting a new venture is to develop a business plan. A business plan is a "road map" to guide the future of the business or venture. The elements of the business plan will have an impact on daily decisions and provide direction for expansion, diversification, and future evaluation of the business. This chapter will assist in drafting a business plan. It includes a discussion of the makeup of the plan and the information that is needed to develop a business plan. Business plans are traditionally developed and written by the owner with input from family members and members of the business team. Business plans are "living" documents that should be reviewed and updated every year or if an opportunity for change presents itself. Reviews reinforce the thoughts and plans of the owner and the business, and aid in the evaluation process. For an established venture, evaluation determines if the business is in need of change or if it is meeting the expectations of the owners.

Importance of a business plan

Planning is very important if a business is to survive. By taking an objective look at a business, areas of weakness and strength can be identified. In planning for a business, needs that may have been overlooked are spotted and nipped in the bud before they escalate. Business plans are established to meet business goals. The business plan is only useful if they are applied. Many new businesses fail in their infancy. Failure is often attributed to a lack of planning.

The business plan is used to enhance the success of a business. A comprehensive, well-constructed business plan can prevent a business from a downward spiral. A business plan provides the information needed to communicate with others; this is especially true if a business is seeking financing. A thorough business plan will have the information to serve as a financial proposal and should be accepted by most lenders. There are six major reasons why a business should have a business plan. These are:

- The process of putting a business plan together forces the person preparing the plan to look at the business in an objective and critical manner.
- It helps to focus ideas and serves as a feasibility study of the business's chances for success and growth.
- The finished report serves as an operational tool to define the company's present status and future possibilities.
- It can help manage the business and prepare it for success.
- It is a strong communication tool for a business. It defines the purpose, competition, management, and personnel of the business. The process of constructing a business plan can be a strong reality check.
- The finished business plan provides the basis for business financing proposal.

Authorship of a business plan

The owner of the business should write the plan. It doesn't matter if the owner is using the business plan to seek financial resources or to evaluate future growth, define a mission, or provide guidance for running the business – the owner knows the most about the business. Consultants can be hired to assist in the process of formulating a business plan, but in reality, the owner of the business must do a majority of the work. Only the owner can come up with the financial data, the purpose of the business, the key employees, management styles, etc.

Components of a business plan

Business plans are critical to the success of any new venture. Entrepreneurs dedicate time to create them, regardless if they are searching for investors. Business plans serve as the framework for a company and provide benchmarks to see if they are reaching their goals. A business plan is key to keeping a business on track. It's important to outline a business plan carefully. All variables should be considered when creating a business pan so as not to test any assumptions. Owners of businesses should work closely with mentors, business partners, and colleagues on creating business plans. Seeking inputs is a great way to get an objective view.

The size and scope of a business plan depend on an organization's specific goals. If the business plan is being drafted for investors, the plan should be very detailed, as potential investors might not be as familiar with the industry as the owner of the business. All concepts have to be explained and where in the business they fit in. If a product or service of the organization is not overly complex, the business plan doesn't have to be very lengthy. A business plan should be structured like a book with the title or cover page first, followed by a table of contents. Following these two pages, the main parts of the plan normally appear in this order: executive summary, business mission statement, goals and objectives, background information, organizational matters, marketing plan, and financial plan. Although the exact structure of a business plan varies, the requirements for plans include the following components:

- **The Executive Summary**

The first page of a business plan should be a persuasive summary that will entice a reader to take the plan seriously and read on. The executive summary is placed at the front of the business plan, but it should be the last part written. The summary describes the proposed business or changes to the existing business and the sector of which the business is (or will be) a part. Research findings and recommendations should be summarized concisely to provide the reader with the information required to make any decisions. The summary outlines the direction and future plans or goals of the business, as well as the methods that will be used to achieve these goals.

The summary should include adequate background information to support these recommendations. The final financial analysis and the assumptions used are also a part of the executive summary. The analysis should show how proposed changes will ensure the sustainability of the current or proposed business. All challenges facing the existing business or proposed venture should also be discussed in this section. Identifying such challenges shows the reader that all considerations have been explored and considered during the research process.

The executive summary should follow the cover page and not exceed two pages in length. The executive summary tells what the company is all about and why it will be successful. It also includes the company's mission statement, product or service, and basic information about the company's leadership team, employees, and location. It should also include financial information and high-level growth plans if the company is asking for financing. The summary should include:

- A brief description of the company's history
- The company's objectives
- A brief description of the company's products or services
- The market the business will compete in
- A persuasive statement as to why and how the business will succeed, discussing the business's competitive advantage
- Projected growth for the company and the market
- A brief description of the key management team
- A description of funding requirements, including a time line and how the funds will be used

- **Mission, Goals, and Objectives**

This section has three separate portions. It begins with a brief, general description of the existing or planned business. The overview is followed by the mission statement of the business. The mission statement consists of key ideas about why the business exists. The third (and final) portion sets the business's goals and objectives. There are at least two schools of thought about goals and objectives. One is that the goals are the means of achieving the objectives, and the other is exactly the opposite – that the objectives are the means of achieving the goals. These goals and objectives should show what the business wishes to accomplish and the steps needed to obtain the desired results. Goals or objectives should follow the acronym SMART, which stands for Specific, Measurable, Attainable, Reasonable, and Timed, to allow for evaluation of the entire process and provide valuable feedback along the way. The business owner should continually evaluate the outcomes of decisions and practices to determine if the goals or objectives are being met and make modifications when needed.

• **Background Information**

Background information should come from the research conducted during
the writing process. This portion should include information regarding the
history of the industry, the current state of the industry, and information
from reputable sources concerning the future of the industry. This portion
of the business plan requires the most investment of time by the writer, with
information gathered from multiple sources to prevent bias or undue opti-
mism. The writer should take all aspects of the industry (past, present, and
future) and business into account. If there are concerns or questions about
the viability of the industry or business, these must be addressed. In writing
this portion of the plan, information may be obtained from the local public library, periodicals, industry personnel, trusted sources on the internet.
Industry periodicals are another excellent source of up-to-date information.
The more varied the sources, the better the evaluation of the industry and
the business, and the greater the opportunity to have an accurate plan.

The business owner must first choose an appropriate legal structure
for the business. The business structure will have an impact on the future,
including potential expansion and exit from the business. If the proper
legal structure is not chosen, the business may be negatively impacted
down the road. Only after the decision is made about the type of business
can the detailed planning begin.

• **The Product or Service**

It is important for the reader to thoroughly understand product offering
or the services currently provided by the company or plan on providing.
However, it is important to explain this section in layman's terms to avoid
confusion. Do not overwhelm the reader with technical explanations or
industry jargon that he or she will not be familiar with. It is important
to discuss the competitive advantage the company's product or service
has over the competition. Or, if the company is entering a new market, it
should answer why there is a need for the new business offering. If appropriate, discuss any patents, copyrights, and trademarks the company currently owns or has recently applied for, and discuss any confidential and
non-disclosure protection the company has secured. Discuss any barriers
that the company faces in bringing the product to market, such as government regulations, competing products, high product development costs,
the need for manufacturing materials, etc. Areas that should be covered in
this section include:

• Is product or service already on the market or is it still in the research
and development stage?

- If still in the development stage, what is the roll out strategy or time line to bring the product to market?
- What makes the product or service unique? What competitive advantage does the product or service have over its competition?
- Can the product or service be sold at higher prices than other competitors and still be competitive and maintain a healthy profit margin?

- **The Market**

This section of the plan is extremely important, because if there is no need or desire for the company's product or service there won't be any customers. If a business has no customers, there is no business. Investors look for management teams with a thorough knowledge of their target market. If the company is launching a new product, include marketing research data. If there are existing customers, provide an analysis of who the customers are, their purchasing habits, and their buying cycle. A good understanding of the industry outlook and target market would be needed. Competitive research will show what other businesses are doing and what their strengths are. In the market research, trends and themes would be required. What do successful competitors do? Why does it work? Can they do it better? This section of the plan should include:

- A general description of the market.
- The niche the business plans on capitalizing on and why.
- The size of the niche market. Include supporting documentation.
- A statement and supporting documentation as to why the business believes there is a need for the product or offering by this market.
- What percentage of the market can the business project capture?
- What is the growth potential of the market? Include supporting documentation.
- Will the business share the market increase or decrease as the market grows?
- How will the business satisfy the growth of the market?
- How will the business price its goods or services in the growing competitive market?

- **The Marketing Strategy**

Once the market has been identified, the company will need to explain a strategy for reaching the market and distributing their product or service. Potential investors will look at this section carefully to make sure there is a viable method to reach the target market identified at a price point that makes sense. Analyze competitors' marketing strategies to learn how they reach the market. If their strategy is working, consider adopting a similar plan. If there is room for improvement, work on creating an innovative plan that will position the company's product or service in the minds of potential customers. The most effective marketing strategies typically integrate multiple mediums or promotional strategies to reach the market. The following are some promotional options to consider:

- TV
- Radio
- Print
- Web
- Direct mail
- Trade shows
- Public relations
- Promotional materials
- Telephone sales
- One-on-one sales
- Strategic alliances

If there are current samples of marketing materials or strategies that have proved successful, they are included in the plan. Developing an innovative marketing plan is critical to a company's success. Investors look favorably upon creative strategies that will put the product or service in front of potential customers. Once it has been identified how to reach the market, discuss in detail strategy for distributing the product or service to customers. The following methods could be used: mail order, personal delivery, hiring of sales representatives, contact with distributors or resellers, etc.

Every purchase decision that a consumer makes is influenced by the marketing strategy or plan of the company selling the product or service. Products are usually purchased based on consumer preferences, including brand name, price, and perceived quality attributes. Consumer preferences develop (and change) over time, and an effective marketing plan takes these preferences into account. This makes the marketing plan an important part of the overall business plan. In order to be viable, the marketing plan must coincide with production activities. The marketing plan must address consumer desires and needs.

A complete marketing plan should identify target customers, including where they live, work, and purchase the product or service the business is providing. Products may be sold directly to the consumer (retail) or through another business (wholesale). Whichever marketing avenue is chosen, when starting a new enterprise or expanding on an existing one it needs to be decided if the market can bear more of what is planned to be produced. The plan must also address the challenges of the marketing strategy proposed. This portion of the plan contains a description of the characteristics and advantages of the product or service. Other variables to consider are sales location, market location, promotion and advertising, pricing, staffing, and the costs associated with all of these. All of these aspects of the marketing plan will take time to develop and should not be taken lightly.

An adequate way of determining the answers to business and marketing issues is to conduct a SWOT analysis. The acronym SWOT stands for Strengths, Weaknesses, Opportunities, and Threats. Strengths represent internal attributes and may include aspects like previous experience in the business. Experience in sales or marketing would be an area of strength for a firm. Weaknesses are also internal and may include aspects such as the time, cost, and effort needed to introduce a new product or service to the marketplace. Opportunities are external aspects that will help the business take off and be sustained. If no business is offering identical products or services in an immediate area, there might be an opportunity to capture the market. Threats are external and may include aspects like other businesses offering the same product in close proximity to the business or government regulations impacting business practices and costs.

- **The Competition**

Analyzing competitors should be an ongoing practice. Knowing competition will allow the company to become more motivated to succeed, becoming more efficient and effective in the marketplace. Understanding the competition's strengths and weaknesses is critical for establishing a product's or service's competitive advantage. If a competitor is struggling, it needs to be known why, so the same mistakes are not made. If competitors are highly successful, the reasons for this have to be identified. The entrepreneur will also want to explain why there is room for another player in the market. Specific areas to address in this section are:

- Identify closest competitors. Where are they located? What are their revenues? How long have they been in business?
- What percentage of the market do they currently have?
- Define their target market.
- How do company's operations differ from competition? What do they do well? Where is there room for improvement?
- In what ways is the business superior to the competition?
- How is their business doing? Is it growing? Is it scaling back?
- How are their operations similar to the business and how do they differ?

- Are there certain areas of the business where the competition surpasses the business? If so, what are those areas and how does the business plan on compensating?

• Operations

Now that there is an opportunity to sell an idea and get potential investors, the next step is how to implement the idea. What resources and processes are necessary to get the product to market? This section of the plan should describe the manufacturing, research and development, purchasing, staffing, equipment, and facilities required for the business. There needs to be a roll out strategy as to when these requirements need to be purchased and implemented. Company financials should reflect the roll out plan. In addition, there should be a description of the vendors that will need to build the business. Questions need to be asked, such as: "Do you have current relationships, or do you need to establish new ones?" "Who will you choose and why?"

• Organizational Matters

This section of the plan describes the current or planned business structure, the management team, and risk management strategies. There are several forms of business structure to choose from including sole proprietorship, partnership, corporations, cooperative, and limited liability corporations or partnerships (LLC or LLP). The type of business structure is an important decision and often requires the advice of an attorney (and an accountant). The business structure should fit the management skills and style(s) of the owner (or owners) and take into account the risk management needs (both liability and financial) of the business. For example, if there is more than one owner (or multiple investors), a sole proprietorship is not an option because more than one person has invested time and/or money into the business. In this case, a partnership, cooperative, corporation, LLC, or LLP would be the proper choice.

If the business is not a sole proprietorship, the management team should be described in the business plan. The management team should consist of all parties involved in the decisions and activities of the business. The strengths and background of management team members should be discussed to highlight the positive aspects of the team. Even if the business is a sole proprietorship, usually more than one person (often a spouse, child, relative, or other trusted person) will have input into the decisions and therefore should be included as team member(s).

Regardless of the business structure, all businesses should also have an external management support team. This external team should consist of the business's lawyer, accountant, insurance agent or broker, and possibly a mentor. These external members are an integral part of the management team. Many large businesses have these experts on staff. For small businesses, the external management team replaces full-time experts; the business owner(s) should consult with this external team on a regular basis (at least once a year) to determine if the business is complying with all rules and regulations. Listing the management team in the business plan allows the reader to know that the business owner has developed a network of experts to provide advice.

The business structure will also determine a portion of the risk management strategy since the way that a business is structured carries varying levels of risk to the owner and/or owners. All marketing strategies (or objectives) carry a degree of risk and must be evaluated, and mitigation strategies should be included in this portion of the plan.

For most investors the experience and quality of the management team is the most important aspect they evaluate when investing in a company. Investors must feel confident that the management team knows its market and product, and has the ability to implement the plan. The plan must communicate management's capabilities in obtaining the objectives outlined in the plan. If this area is lacking, chances for obtaining financing are bleak. If the management team lacks in a critical area, identify how to plan on compensating for the void. Whether it is additional training required or additional management staff needed, show that the problem exists, and provide options for solutions. The following areas should be addressed when preparing this section of the business plan:

Personal history of the principals:

- Business background of the principals
- Past experience – tracking successes, responsibilities, and capabilities
- Educational background (formal and informal)
- Personal data: age, current address, past addresses, interests, education, special abilities, reasons for entering into a business
- Personal financial statement with supporting documentation

Work experience:

- Direct operational and managerial experience in this type of business
- Indirect managerial experiences

Duties and responsibilities:

- Who will do what and why?
- Organizational chart with chain of command and listing of duties
- Who is responsible for the final decisions?

Salaries and benefits:

- A simple statement of what management will be paid by position
- Listing of bonuses in realistic terms
- Benefits (medical, life insurance, disability. . .)

Resources available to the business:

- Insurance broker(s)
- Lawyer
- Accountant
- Consulting group(s)
- Chambers of Commerce
- Local colleges and universities
- Federal, state, and local agencies
- Board of Directors
- World Wide Web (various search engines)
- Banker

- **Personnel**

The success of a business can often be measured by its employees. The majority of consumers will go elsewhere if they don't receive prompt and courteous service. The following factors must be considered when completing this section of the business plan:

- What are the current company's personnel needs? How many employees are envisioned in the near future and then in the next three to five years?
- What skills must employees have? What will their job descriptions be?
- Are the people needed readily available and how will they be attracted to the company?
- How will salaries be paid?
- Will there be benefits? If so, what will they be and at what cost?
- Will overtime be paid?

Financial information

At the heart of any business operation is the accounting system. It is important to have a certified public accountant establish the company's accounting system before the start of business. At times there is a tendency to personally do accounts. An incredible number of businesses fail due to managerial inefficiencies. It's better for trained professionals to help in the area of accounting and legal matters. If the business can't afford a public accountant to establish accounting procedure, then the company is undercapitalized. The business needs to secure additional resources before starting.

The financial plan and assumptions are crucial to the success of the business and should be included in the business plan. One of the foremost reasons new businesses fail is not having enough start-up capital or inadequate planning to cover all expenses and be profitable. The scope of the business will be determined by the financial resources that can be acquired. Because of this, a financial plan will need to be developed and the supporting documents to substantiate it need to be created. The financial plan has its basis in historical data (for an existing business) or from projections (for a proposed business). The first issue to address is record-keeping. Who will keep the necessary records should be indicated and how these records will be used. Internal controls, such as who will sign checks and handle any funds, should also be addressed in this section. A good rule to follow for businesses other than sole proprietorships is having at least two people sign all checks.

The next portion of the financial plan should be assumptions concerning the source of financing. This includes if (and when) the business will need additional capital: How much capital will be needed and how will these funds be obtained? If start-up capital is needed, this information should be included in this portion. Personal contributions should be included along with other funding sources. The amount of money and repayment terms should be listed. One common mistake affecting many new businesses is underfunding at start-up. Owners too often do not carefully evaluate all areas of expense and underestimate the amount of capital needed to see a new business through the development stages (including living expenses, if off-firm income is not available).

One of the first steps to having a profitable business is to establish a bookkeeping system that provides the business with data. Typically, a balance sheet, income statement, cash flow statement, and partial budget or enterprise budgets are included in a business plan. These documents will display the financial information in a form that lending institutions are used to seeing. If these are not prepared by an accountant, having one review them will ensure that the proper format has been used. Financial projections should be completed for at least two years and, ideally, for five years.

Balance Sheet – This indicates what the cash position of the business is and what the owner's equity is at a given point (the balance sheet will show assets, liabilities, and retained earnings). A balance sheet is a snapshot of a business's assets and liabilities and its owner's equity at a specific point in time. A balance sheet can be prepared at any time, but is usually done at the end of the fiscal year (for many businesses, this is the end of the calendar year). Evaluating the business by using the balance sheet requires several years of balance sheets to tell the true story of the business's progress over time. A balance sheet is typically constructed by listing assets on the left and liabilities and owner's equity on the right. The difference between the assets and liabilities of the business is called the "owner's equity" and provides an estimate of how much of the business is owned outright. Owner's equity provides the "balance" in a balance sheet.

Assets are anything owned by or owed to the business. These include cash (and checking account balances), accounts receivable (money owed to the business), inventory (any inventory that the business has), land, equipment, and buildings. This may also include machinery, breeding stock, small fruit bushes or canes, and fruit trees. Sometimes assets are listed as current (those easily converted to cash) and fixed (those that are required for the business to continue). Assets are basically anything of value to the business.

Balance sheets may use a market basis or a cost basis to calculate the value of assets. A market-basis balance sheet better reflects the current economic conditions because it relies on current or market value for the assets rather than what those assets originally cost. Market values are more difficult to obtain because of the difficulty in finding accurate current prices of assets and often results in the inflation of the value of assets. Cost-basis balance sheets are more conservative because the values are often from prior years. For example, a cost-basis balance sheet would use the original purchase price of land rather than what selling that land would bring today. Because purchase records are easily obtained, constructing a cost-basis balance sheet is easier. Depreciable assets, such as buildings, tractors, and equipment, are listed on the cost-basis balance sheet at purchase price less accumulated depreciation. Most accountants use the cost-basis balance sheet method.

Liabilities are what the business owes on the date the balance sheet is prepared. Liabilities include both current liabilities (accounts payable, any account the business has with a supplier, short-term notes, operating loans, and the current portion of long-term debt, which are payable within the current year) and non-current liabilities (mortgages and loans with a term that extends over one year).

Owner's equity is what remains after all liabilities have been subtracted from all assets. It represents money that the owner has invested in the business, profits that are retained in the business, and changes caused by fluctuating market values (on a market-basis balance sheet). Owner's equity will be affected whenever changes in capital contributed to the business or there are retained earnings; so, if the business practice is to use all earnings as the "paycheck" rather than reinvesting them in the business, the owner's equity will be impacted. On the balance sheet, owner's equity plus liabilities equals assets. Or, stated another way, all of the assets less the amount owed (liabilities) equals the owner's equity (sometimes referred to as "net worth").

Income Statement – also called the profit and loss statement, is used to indicate how well the company is managing its cash, by subtracting disbursements from receipts. The income statement is a summary of the income (revenue) and expenses for a given accounting cycle. If the balance sheet is a "snapshot" of the financial health of the business, the income statement is a "motion picture" of the financial health of the business over a specific time period. An income statement is constructed by listing the income (or revenue) at the top of the page and the expenses (and the resulting profit or loss) at the bottom of the page.

Revenue is any income realized from sales, government payments, and any other income the business may have (including such items as fuel tax refunds, patronage dividends, and custom work). Other items affecting revenues are changes in inventory and accounts receivable between the start of the time period and the end, even if these changes are negative. Expenses include any expense the business has incurred from production or products sold. Depreciation, which is calculated wear and tear on assets (excluding land), is included as an expense for accounting purposes. Interest is considered an expense, but any principal payments related to loans are not an expense.

As the income statement is created, the desired outcome is to have more income than expenses, so the income statement shows a profit. If not, the final number is shown in parentheses (signifying a negative number). Another name for this financial record is a "profit and loss statement." Income statements are one way to clearly show how the firm is making progress from one year to the next and may provide a much more optimistic view of sustainability than can be seen by looking at a single year's balance sheet.

Cash Flow – this projects all cash receipts and disbursements. Cash flow is critical to the survival of any business. A cash flow statement is the predicted flow of cash into and out of a business over a year. Cash flow statements are prepared by showing the total amounts predicted for each item of income or expense. This total is then broken down by month to show when surpluses and shortfalls in cash will occur. In this way, the cash flow statement can be used to predict when additional cash is needed and when the business will have a surplus to pay back any debt. This monthly prediction allows the owner(s) to better evaluate the cash needs of the business, taking out applicable loans and repaying outstanding debts. The cash flow statement often uses the same categories as the income statement plus additional categories to cover debt payments and borrowing.

Break-even Analysis – this is based on the income statement and cash flow. All businesses should perform this analysis without exception. A break-even analysis shows the volume of revenue from sales that are needed to balance the fixed and variable expenses.

The break-even analysis calculates the point where the business has reached a zero balance (i.e., when income covers expenses exactly). Before the break-even point is calculated, the following details should be known:

- Timeframe (e.g., monthly/yearly)
- Average price of each product/service sold
- Average cost of each product/service to make/deliver
- Fixed costs for the month/year

Once the above figures are known, the break-even can be worked out by completing the calculations below:

- Percentage of price that is profit – Calculate (average price of each product/service sold minus average cost of each product/service to make/deliver) divided by average price of each product/service sold.
- Number of units sold needed to break-even – Calculate fixed costs for the month/year divided by (average price of each product/service sold minus average cost of each product/service to make/deliver).
- Total sales needed to break-even – Calculate number of units sold needed to break-even multiplied by average price of each product/ service sold.

After these financial statements are completed, the business plan writer will have an accurate picture of how the business has performed and can project how the business will perform in the coming year(s). With such information, the owner and any readers of the business plan will be able to evaluate the viability of the business and have an accurate understanding of actions and activities that will contribute to its sustainability. This understanding will enable the owner(s) to make better informed decisions regarding loans or investments in the business.

If the goal of the business plan is to obtain financing, generating financial forecasts will be required in the plan. The forecasts demonstrate the need for funds and the future value of equity investment or debt repayments. This exercise is critical in obtaining capital for the business. To obtain capital from lending institutions, it must be demonstrated that there is the need for the funding and the ability to repay the loan. The forecast generated should cover a three- to five-year period. This is a period in which realistic goals can be established and attained without much speculation. Forecasts should be broken down in monthly increments.

Supplement the funding request with financial projections. The goal is to convince the reader that the business is stable and will be a financial success. If the business is already established, income statements, balance sheets, and cash flow statements for the last three to five years should be included. If there are other collateral to put against a loan, they should be listed. Forecasted income statements, balance sheets, cash flow statements, and capital expenditure budgets should be included. Projections should be clearly explained and matched with funding requests. Graphs and charts can be applied to tell the financial history of the business.

After the mission, background information, organization, and marketing and financial plans are complete, an executive summary can then be prepared. Armed with the research results and information from the other sections, the business will then come alive. The next step is to share this plan with others whose opinions are respected and needed. Have them ask the hard questions: making the business defend an opinion expressed or challenging the business to describe what they plan to do in more detail. Often people are hesitant to share what they have written with their families or friends because they fear the plan will not be taken seriously. However, it is much better to receive constructive criticisms from family and friends (and gain the opportunity to strengthen the business plan) than it is to take it immediately to the lender, only to have any problems pointed out and receive a rejection.

The presentation of the plan should be as professional as possible to portray the business in a positive manner. When dealing with a lender or possible investor, the plan will be reviewed for accuracy and suggestions for changes to the plan may be offered. The decision to recommend the loan to the appropriate committee or reject the proposal will be largely based on the business plan. Often loan officers will not know a great deal about the proposed venture, but they will know the correct structure of a business plan. Investors will make their decision based on the plan and the integrity of the owner. For this reason, it is necessary to use a professional format. After loan officers complete their evaluations, the loan committee will further review the business plan and make a decision. The committee members will often spend limited time reviewing the document, focusing on the message of the executive summary and financial statements to make their determination. Because of this, these portions need to be the strongest parts of the plan and based on sound in-depth research and analysis.

Once all parts of the business plan have been written, it provides a document that will enable the business to do analysis and determine which, if any, changes need to be made. Changes on paper take time and effort but are not as expensive as changing a business practice only to find that the chosen method is not viable. For a proposed venture, if the written plan points to the business not being viable, large sums of money have not been invested and possibly lost. In short, challenges are better faced on paper than with investment capital.

Conclusion

It's important to keep in mind that major events in a business's target marketplace (e.g., competitor consolidation, acquisition of a major customer) or in the broader environment (e.g., new legislation) could trigger a review of a business's strategic objectives. Regardless of whether or not there are fixed time intervals in a business plan, it must be part of a rolling process with regular assessment of performance against the plan and agreement of a revised forecast if necessary.

For most businesses, an annual business plan, broken down into four quarterly business operating plans, is required. However, if a business is heavily dependent on sales it can make more sense to have a monthly business operating plan, supplemented where necessary with weekly targets and reviews. Targets make it clearer for individual employees to see where they fit within an organization, and what they need to do to help the business meet its objectives. Setting clear objectives and targets and closely monitoring their delivery can make the development of a business more effective. Targets and objectives should also form a key part of employee appraisals, as a means of objectively addressing individuals' progress.

Bibliography

Adams, B. (1998). *Adams Streetwise Complete Business Plan*, Adams Media, New York.

Markides, C. (1999). *All the Right Moves: A Guide to Drafting Breakthrough Strategy*, Harvard Business School Press, Brighton, MA.

Pinson, J. (1999). *Anatomy of a Business Plan*, Dearborn Trade, Chicago.

Press, E. (1999). *Analyzing Financial Statements*, Lebhar-Friedman, New York.

Schwartz, P. (1996). *The Art of the Long View*, Currency-Doubleday, New York.

chapter eight

Business modeling and forecasting

Introduction

The aim of this chapter is to support modeling on an enterprise scale. This chapter brings about modeling and decision automation to improve the quality and reliability of data for efficient decision-making. It consists of advanced tools and expert systems to sharpen business operations and contribute to competitive advancement. Modeling is integral to business management. Models are frameworks to assess competing courses of action. They help compare risk-reward scenarios and test ideas. Modeling is also pervasive. Every time spreadsheets are built to run the numbers, models are being built. Modeling is at the core of enterprise strategy and treating it as such is nothing less than imperative.

Applying modeling techniques on an enterprise scale is frequently called Enterprise Simulation (ES). It has two distinct objectives:

- Provide a top-down view of the business enterprise to support strategic decision-making.
- Enable workers across the enterprise with modeling tools and techniques for routine decisions.

Business Intelligence (BI) and similar historical reporting systems help monitor business operations, but they cannot evaluate new ideas for which there is no history. Spreadsheets are flexible but unable to scale or foster collaboration. Business Analytics modeling is unique in its ability to satisfy ES objectives, cut modeling costs by simplifying model structures and encouraging reuse of prior work, and to improve model accuracy by dividing tasks among those with the best domain knowledge.

The ability to model and perform decision modeling and analysis is an essential feature of many real-world applications ranging from emergency medical treatment in intensive care units to military command and control systems. The analysis and modeling of real-world problems are important features for performing decision-making in diverse applications. This range from manufacturing applications, construction, military applications, health applications, logistic, transportation distribution, to mention just a few. Models are essential in providing support for businesses processes and systems and dealing with complex problems. The development of appropriate models for planning and management is a tool for improving efficiency in real-world problems. Consequently, models are developed, with the aim of computing estimates and forecasts for real-world data. Modeling helps to make informed decisions, using techniques for analysis, estimation, and forecasting. With modeling, data can be used to describe realities, build scenarios, and predict performances.

Almost all decision-making is based on forecasts. Every decision becomes operational at some point in the future, so it should be based on forecasts of future conditions. Forecasts are needed continually, and as time moves on, the impact of the forecasts on actual performance is measured, original forecasts are updated, decisions are modified, and so on. If models are accurate, reliable forecasts can be computed based on the models. The resulting forecasts yield the same performances as that of the models.

Almost all managerial decisions are based on forecasts. Every decision becomes operational at some point in the future, so it should be based on forecasts of future conditions. Forecasts are needed continually, and as time moves on, the impact of the forecasts on actual performance is measured, original forecasts are updated, and decisions are modified, and so on. The decision-maker uses forecasting models to assist him or her in the decision-making process. The decision-making often uses the modeling process to investigate the impact of different actions under a course of action.

Modeling theory

A model is defined as a representation of a system for the purpose of studying the system. A model is a representation and abstraction of anything such as a real system, a proposed system, a futuristic system design, an entity, a phenomenon, or an idea. It is necessary to consider only those aspects of the system that affect the problem under investigation. These aspects are represented in a model, and by definition it is a simplification of the system. It is possible to derive a model based on physical laws, which it is possible to calculate the value of some time-dependent quantity nearly exact at any instant of time. If exact calculations were possible, such a model would be entirely *deterministic*. Nevertheless, it may be possible to derive a model that can be used to calculate the probability of a future behavior of the value lying between two specified limits. Such a model is called a probability model or a *stochastic model*.

The external components which interact with the system and produce necessary changes are said to constitute the system environment. In modeling systems, it is necessary to decide on the boundary between the system and its environment. This decision may depend on the purpose of the study. A model is closely related to the system. A system is defined as an aggregation or assemblage of objects joined in some regular inter-action or interdependence toward the accomplishment of some purpose. The system environment constitutes external components which interact with the system and produce necessary changes to the system. In model-ing systems, it is necessary to decide on the boundary between the system and its environment. This decision may depend on the purpose of the study. The components of a system are defined as follows:

Entity: An entity is an object of interest in a system. For example, in a production system, departments, orders, parts, and products are the entities.

Attribute: An attribute denotes the property of an entity. For example, quantities for each order, types of parts, or number of machines in a department are attributes of a production system.

Activity: Any process causing changes in a system is called an activity. For example, the fabrication process of a department.

State of a system: The state of a system is defined as the collection of variables necessary to describe a system at any time, relative to the objective of the study. In other words, the state of the system means a description of all the entities, attributes, and activities as they exist at one point in time.

Example

An example of a system is the production system as illustrated in Figure 8.1. As shown in this figure, there are certain distinct objects, each of which possesses properties of interest. There are also certain interac-tions occurring in the system that cause changes in the system.

In a production system, the arrival of orders may be considered to be outside the company but yet a part of the system environment. There is a relationship between company output and arrival of orders in considering

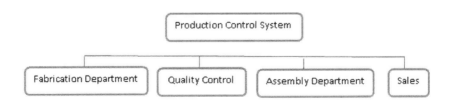

Figure 8.1 Production system.

demand and supply of goods. This relationship is considered an activity of the system.

Every study begins with a statement of the problem, provided by policymakers. Analysts ensure it's clearly understood. This step is referred to as problem formulation. Another step of the modeling process is the setting up of objectives. The objectives indicate the questions to be answered by identifiable problems. The construction of a model of a system or model conceptualization is an art as science. The modeling process is characterized by the ability (1) to abstract the essential features of a problem (2) to select and modify basic assumptions that characterize the system, and (3) to enrich and elaborate the model until useful approximation results are obtained. The model building process enhances the quality of the resulting model and increases the confidence of the model user in the application of the model. Real-world systems result in models that require a great deal of information storage and computation. These models can be programmed using system languages or special purpose software. Examples of these software include MATLAB, Modeller, MacA&D, ArchCAD, AutoCAD, CATIA, etc. The programming of these models is referred to as the model translation.

It is pertinent to verify and validate the models by checking the performance of the models developed to represent real-world situations. If the input parameters and logical structure are correctly represented, verification is completed. As part of the calibration process of a model, the modeler must validate and verify the model. The term validation is applied to those processes, which seek to determine whether or not a model is correct with respect to the "real" system. More prosaically, validation is concerned with the question "Are we building the right system?" Verification, on the other hand, seeks to answer the question "Are we building the system right?" The purpose of modeling is to determine that a model is an accurate representation of the real system. This is achieved through calibration of the model, which is an iterative process of comparing the model to actual system behavior and the discrepancies between the two are determined.

The model obtained for a problem can be generic, i.e., can be used again by the same or different analysts to solve different problems. Further modifications can be made. Model users can change the input parameters for better performance. Success of the model depends on if model requirements and outputs are fully implemented. The purpose of models is to aid in designing solutions. They are to assist in understanding the problem and to aid deliberation and choice by allowing us to evaluate the consequence of our action before implementing them. The modeling process is given in Figure 8.2.

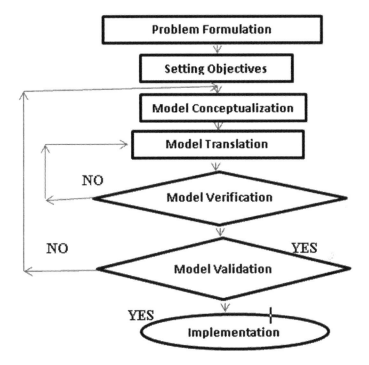

Figure 8.2 Modeling process.

Types of models

In modeling theory, there are many models that are designed but few of them are used. In formulating models, the concept of "implementation" is defined, and we progressively shift from a traditional "design then implementation" standpoint to a more general theory of a model design/implementation; seen as a cross-construction process between the model and the organization in which it is implemented. Consequently, the organization is considered not a simple context, but as an active component in the design of models. In model-based decision-making, we are particularly interested in the idea that a model is designed with a view to action. Basically, there are various types of models. They are:

> *Mathematical Model* – This is one in which symbols and logic constitute the model. The symbolism used can be language or mathematical notations. This type of model uses mathematical equations to represent a system.
>
> **Example:** A typical example of a mathematical model is a simulation model. This is the study of the behavior of a system as it

evolves over time. This model takes the form of a set of assumptions concerning the operation of the system. These assumptions are expressed in the form of

- Mathematical relationships
- Logical relationships
- Symbolic relationships

Physical (Iconic) Model – The physical model is a smaller or larger physical copy of an object.

The object being modeled may be small (for example, an atom) or large (for example, the solar system). Other examples include a map, a globe, or a model car.

Static Model – This method is also known as the Monte-Carlo method. This system represents a system at a particular point in time. It describes relationships that do not change with respect to time. Examples of this type of model include an architectural model of a house, or an equation relating the lengths and widths on each side of a playground variation.

Dynamic Model – This type of model represents systems as they change over time. Dynamic system means a system capable of action and/or change. The dynamic model describes time-varying relationships. Examples include a wind tunnel, or equations of the motion of the planets around the sun that constitute a dynamic model of the solar system.

Deterministic Model – This type of models contains no random variables. They have a known set of inputs which will result in a unique set of outputs. An example is the arrival of patients to a hospital at their scheduled appointment time.

Stochastic Model – The stochastic model has one or more random variables as inputs. These random inputs lead to random outputs. For example, a banking system involves random inter-arrival and service times.

Discrete Model – This is the discrete analogue of continuous modeling. In discrete models, formulae are fit to discrete data. Discrete data are data that could potentially take only a countable set of values. These could be integers and are not infinitely divisible. An example is the number of cells in a population, displayed at regular time intervals.

Continuous Model – This is the mathematical practice of applying a model to continuous data (data which has an infinite number and divisibility of attributes). They often are in the form of differential equations and are converted to discrete models. An example is the amount of water in a tank and or its temperature.

There are guidelines for an analyst to successfully implement a model that could be both valid and legitimate. These guidelines are as follows:

1. The analyst must be ready to work in close cooperation with the strategic stakeholders in order to acquire a sound understanding of the organizational context. In addition, the analyst should constantly try to discern the kernel of organizational values from its more contingent part.
2. The analyst should attempt to strike a balance between the level of model sophistication/complexity and the competence level of stakeholders. The model must be adapted both to the task at hand and to the cognitive capacity of the stakeholders.
3. The analyst should attempt to become familiar with the various preferences prevailing in the organization. This is important since the interpretation and the use of the model will vary according to the dominant preferences of the various organizational actors.
4. The analyst should make sure that the possible instrumental uses of the model are well documented and that the strategic stakeholders of the decision-making process are quite knowledgeable about and comfortable with the contents and the working of the model.
5. The analyst should be prepared to modify or develop a new version of the model, or even a completely new model, if needed, that allows an adequate exploration of heretofore unforeseen problem formulation and solution alternatives.
6. The analyst should make sure that the model developed provides a buffer or leaves room for the stakeholders to adjust and readjust themselves to the situation created by the use of the model.
7. The analyst should be aware of the pre-conceived ideas and concepts of the stakeholders regarding problem definition and likely solutions; many decisions in this respect might have been taken implicitly long before they become explicit.

Modeling for forecasting

Forecasting is an important tool that is useful in planning, whether in business or government. Often, forecasts are generated subjectively and at great cost by group discussion, even when relatively simple quantitative methods can perform just as well or, at very least, provide an informed input to such discussions. The modeling process is useful for:

1. Understanding the underlying mechanism generating the time series. This includes describing and explaining any variations, seasonality, trends, etc.
2. Predicting the future
3. Controlling the system

The selection and implementation of proper forecasting models are necessary for planning purposes in organizations. Usually, the financial well-being of an organization relies on the accuracy of the forecast since such information will likely be used to make interrelated budgetary and operative decisions in areas of personnel management, purchasing, marketing and advertising, capital financing, etc. For example, under-forecasts may cause an organization to be overly burdened with excess asset costs, and hence create lost sales revenue through unanticipated item shortages. Figure 8.3 highlights the systematic development of the modeling and forecasting phases.

There are two main approaches to forecasting. Either the estimates of future values are based on the analysis of factors which are believed to influence future values, i.e., the explanatory method, or else the prediction is based on an inferred study of past general data behavior over time, i.e., the extrapolation method. For example, the belief that the sale of doll clothing will increase from current levels because of a recent advertising blitz rather than proximity to Christmas illustrates the difference between the two philosophies. It is possible that both approaches will lead to the creation of accurate and useful forecasts, but it must be remembered that, even for a modest degree of desired accuracy, the former method is often more difficult to implement and validate than the latter approach.

Time series forecasting

Organizations with a large operation and staff comprised of statisticians, management scientists, computer analysts, etc. are in a much better position to select and make proper use of sophisticated forecast techniques than companies with more limited resources. Notably, the bigger firm,

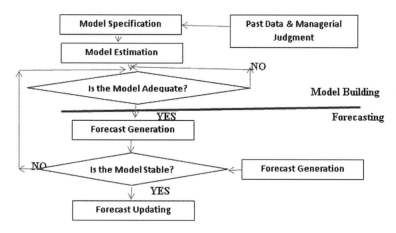

Figure 8.3 Forecasting system: Model-building and forecasting.

through its larger resources, has a competitive edge over smaller organizations and can be expected to be very diligent and detailed in estimating forecasts.

A time series is a sequence of numerical data points in successive order. Generally, a time series is a sequence taken at successive equally spaced points in time. It is a sequence of discrete-time data. For example, in investment analysis, a time series tracks the movement of the chosen data points, such as a security's price over a specified period of time with data points recorded at regular intervals. One of the main goals of time-series analysis is to forecast future values of the series. A trend is a regular, slowly evolving change in the series level. Changes can be modeled by low-order polynomials. There are three general classes of models that can be constructed for purposes of forecasting or policy analysis. Each involves a different degree of model complexity and presumes a different level of comprehension about the processes one is trying to model. In making a forecast, it is also important to provide a measure of how accurate one can expect the forecast to be. The use of intuitive methods usually precludes any quantitative measure of confidence in the resulting forecast. The statistical analysis of the individual relationships that make up a model, and of the model as a whole, makes it possible to attach a measure of confidence to the model's forecasts.

In time-series models, we presume to know nothing about the causality that affects the variable we are trying to forecast. Instead, we examine the past behavior of a time series in order to infer something about its future behavior. The method used to produce a forecast may involve the use of a simple deterministic model such as a linear extrapolation or the use of a complex stochastic model for adaptive forecasting. One example of the use of time-series analysis would be the simple extrapolation of a past trend in predicting population growth. Another example would be the development of a complex linear stochastic model for passenger loads on an airline. Time-series models have been used to forecast the demand for airline capacity, seasonal telephone demand, the movement of short-term interest rates, and other economic variables. Time-series models are particularly useful when little is known about the underlying process one is trying to forecast. The limited structure in time-series models makes them reliable only in the short run, but they are nonetheless rather useful.

In regression models, the variable under study is explained by a single function (linear or non-linear) of a number of explanatory variables. The equation will often be time-dependent (i.e., the time index will appear explicitly in the model), so that one can predict the response over time of the variable under study. The main purpose of constructing regression models is forecasting. A forecast is a quantitative estimate (or set of estimates) about the likelihood of future events which is developed on the basis of past and current information. This information is embodied in the form of a model. This model can be in the form of a single-equation structural model, a multi-equation model or a time-series model. By extrapolating our models beyond the period over which they were estimated, we can make forecasts about near future events.

The term forecasting is often thought to apply solely to problems in which we predict the future. An example of a single-equation regression model would be an equation that relates a particular interest rate, such as the money supply, the rate of inflation, and the rate of change in the gross national product. The choice of the type of model to develop involves trade-offs between time, energy, costs, and desired forecast precision. The construction of a multi-equation simulation model may require large expenditures of time and money. The gains from this effort may include a better understanding of the relationships and structure involved as well as the ability to make a better forecast. However, in some cases these gains may be small enough to be outweighed by the heavy costs involved. Because the multi-equation model necessitates a good deal of knowledge about the process being studied, the construction of such models may be extremely difficult.

The decision to build a time-series model usually occurs when little or nothing is known about the determinants of the variable being studied, when a large number of data points are available, and when the model is to be used mostly for short-term forecasting. Given some information about the processes involved, however, it may be reasonable for a forecaster to construct both types of models and compare their relative performance. Two types of forecasts can be useful. Point forecasts predict a single number in each forecast period, while interval forecasts indicate an interval in which we hope the realized value will lie.

Verification and validation of models

As a means of verification and validation of estimates obtained from a model, the estimates are compared with actual data. A good model should have small error measures between the estimated values and actual data. A model is acceptable if the *mean average percentage error* (MAPE) of the model is less than 20%. This validation technique investigates the performance of newly developed models compared with actual data. The process of computing: MAPE values are outlined below.

The forecast errors are computed from a time series, based on an average of weighted past observations. At period t, past values of a variable of interest X_t can be observed or values forward in the future. The model is applied to the historical observations, and forecasted values F_{t+1} are obtained. To identify an accurate predictive model, the following steps are followed:

- Choose a forecasting method based on the observed pattern of the time series.
- Use the forecasting method to develop fitted values of the data.
- Calculate the forecast error.
- Make a decision about the appropriateness of the model based on the measure of forecast error.

For the purpose of computing forecasting errors, a historical data set called a time series is considered. A time series consists of the data of interest on a single variable which has been collected in a consistent way over time at equally spaced intervals. Time series are analyzed to search for patterns by graphing or plotting time series data set. The one-sided moving average (MA) of past n observations is given as:

$$F_{t+1} = \frac{X_t + X_{t-1} + \ldots + X_{t-n+1}}{n}$$

$$= \frac{1}{n}\left(\sum_{i=t-n+1}^{t} X_i\right)$$

where t is the most recent observation and $t+1$ is the next period. This formula requires that the values of the past n observations are known. Accordingly, the concept of adding a new observation and dropping the oldest observation, the formula is restated as

$$F_{t+1} = \frac{1}{n}\left(\sum_{i=t-n}^{t-1} X_i\right) + \frac{1}{n}\left(X_t - X_{t-n}\right)$$

$$= F_t + \frac{X_t}{n} - \frac{X_{t-n}}{n}$$

The formula representing the time series is an adjustment of the forecast F_t in the previous period. If n is increased, a much smaller adjustment is made for each new time period.

This representation of time series can be developed, by making the substitution $F_t = X_{t-n}^2$, to get

$$F_{t+1} = \frac{X_t}{n} - \frac{F_t}{n} + F_t$$

Furthermore, this can then be rewritten as

$$F_{t+1} = \frac{1}{n}X_t + \left(1 - \frac{1}{n}\right)F_t$$

This is a forecast based on weighting the most recent observations with a weight of value $1/n$ and weighting the most recent forecast with a weight of $1-1/n$. Since the number of periods n is a constant, the fraction $1/n$ must be greater than zero and less than unity. If $\propto$ is substituted for $1/n$, the basic model is written as

$$F_{t+1} = \propto X_t + (1-\propto)F_t$$

where t is the current time period, F_{t+1} and F_t are the forecast values for the next and current periods, and X_t is the current observed value. $\propto$ is called the smoothing constant and it takes values between zero and unity. If F_t is expressed in terms of the preceding observed X_{t-1} and the forecast F_{t-1} values, then the equivalent for the next period's forecast becomes

$$F_{t+1} = \propto X_t + (1-\propto)\left[\propto X_{t-1} + (1-\propto)F_{t-1}\right]$$

which simplifies to

$$F_{t+1} = \propto X_t + \propto (1-\propto)X_{t-1} + (1-\propto)^2 F_{t-1}$$

This can continue for several earlier periods which show that all preceding values of X are reflected in the current forecasts. Thus, the successive weights $\propto, \propto(1-\propto), (1-\propto)^2, \ldots$ decrease exponentially. This can be rewritten as follows:

$$F_{t+1} = F_t + \propto (X_t - F_t)$$

and this is simply

$$F_{t+1} = F_t + \propto e_t$$

where the forecast error e_t for period t, is just the actual minus the forecast. Thus, the forecast provided by the time series is the old forecast plus an adjustment for the error occurring in the last forecast.

To evaluate the accuracy of a predictive technique, it is required to evaluate the error in forecasting. The error in period t was defined as the actual value X_t minus the predicted value F_t:

$$e_t = X_t - F_t$$

An examination of the error in forecasting permits the evaluation of whether the chosen prediction model accurately mirrors the pattern exhibited in the sample observations. An evaluation of the reliability of a model requires the specification of criteria.

The MAPE is based on the assumption that the severity of error is linearly related to its size. The MAPE is the sum of the absolute values of the errors divided by the corresponding observed values divided by the number of forecasts. The MAPE is often expressed as a percentage. This is to provide meaningful interpretation about each data point. The MAPE is defined as:

$$\text{MAPE} = \frac{\sum |e_t| / X_t}{n} \times 100$$

If the predictive model is accurate, reliable forecasts can be computed based on the predictive model. Future demand for power can be determined by developing accurate estimates for predictive models. Predictive models are static, constructed from historical data. The resulting forecasts yield the same performance as that of the predictive model. The flowchart showing the steps for validating results is given in Figure 8.4.

Example

A table showing the actual figures for shoes produced by AZ Company, together with their respective estimated production values is displayed in Table 8.1. The values are given for October 1 to October 23, 2017.

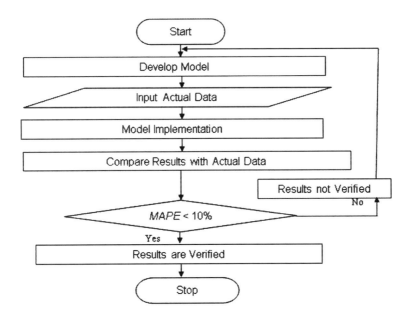

Figure 8.4 Flowchart for validation.

Table 8.1 Actual production for shoes and
their estimated values

Date	Actual	Estimated
Oct 1	39	46
Oct 2	43	48
Oct 3	40	44
Oct 4	42	48
Oct 5	39	46
Oct 6	34	40
Oct 7	39	45
Oct 8	40	44
Oct 9	42	45
Oct 10	41	46
Oct 11	42	46
Oct 12	38	43
Oct 13	35	40
Oct 14	40	45
Oct 15	41	47
Oct 16	42	45
Oct 17	42	46
Oct 18	42	46
Oct 19	36	40
Oct 20	35	39
Oct 21	41	45
Oct 22	42	46
Oct 23	40	44

The MAPE is given as:

$$\text{MAPE} = \frac{\sum |e_t| / X_t}{n} \times 100$$

where,

$$e_t = X_t - F_t$$

The calculations are shown in Table 8.2.

$$\text{MAPE} = \frac{\sum |e_t| / X_t}{n} \times 100 = \frac{2.762923412}{23} \times 100 = 12.0127\%$$

Since the MAPE is less than 20%, the estimated values for the model are acceptable.

Table 8.2 Calculations for MAPE

	Actual	Estimated	e_t	$\lvert e_t \rvert$	$\lvert e_t \rvert / X_t$
t	X_t	F_t	$X_t - F_t$	$\lvert X_t - F_t \rvert$	$\lvert X_t - F_t \rvert / X_t$
1	39	46	−7	7	0.179487
2	43	48	−5	5	0.116279
3	40	44	−4	4	0.1
4	42	48	−6	6	0.142857
5	39	46	−7	7	0.179487
6	34	40	−6	6	0.176471
7	39	45	−6	6	0.153846
8	40	44	−4	4	0.1
9	42	45	−3	3	0.071429
10	41	46	−5	5	0.121951
11	42	46	−4	4	0.095238
12	38	43	−5	5	0.131579
13	35	40	−5	5	0.142857
14	40	45	−5	5	0.125
15	41	47	−6	6	0.146341
16	42	45	−3	3	0.071429
17	42	46	−4	4	0.095238
18	42	46	−4	4	0.095238
19	36	40	−4	4	0.111111
20	35	39	−4	4	0.114286
21	41	45	−4	4	0.097561
22	42	46	−4	4	0.095238
23	40	44	−4	4	0.1
	Total				2.762923412

An array of forecasting techniques

Forecasting is a prediction of what will occur in the future, and it is an uncertain process. Because of the uncertainty, the accuracy of a forecast is as important as the outcome predicted by the forecast. This section presents a general overview of business forecasting techniques as classified in Figure 8.5. The forecasting techniques considered in thus section include regression analysis, Box-Jenkins, artificial neural network (ANN), and Kalman techniques.

Some of the aforementioned techniques can be solved analytically, for example, the regression technique and the MAs. The analytical method is used to solve a specific issue. This is a procedure for the analysis of some

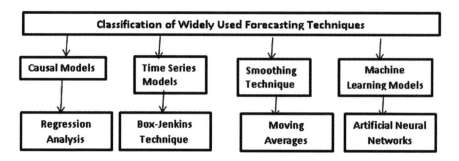

Figure 8.5 General overview of forecasting techniques.

problem or fact. However, some very complex problems are solved using simulation method, for example, the Box-Jenkins technique and the ANN. This is the imitation of a real-world process. The act of simulation requires that a model be developed. The model represents the key characteristics, behaviors, and functions of the process.

Regression analysis

Regression techniques belong to the class of causal models. Regression is the study of relationships among variables, a principal purpose of which is to predict or estimate the value of one variable from known or assumed values of other variables related to it. To make predictions or estimates, we must identify the effective predictors of the variable of interest. To develop a regression model, begin with a hypothesis about how several variables might be related to another variable and the form of the relationship. A regression using only one predictor is called a simple regression. Where there are two or more predictors, multiple regressions analysis is employed.

Given a line $y = mx + b$.

We determine what the best line's slope (m) and intercept (b) should be for a given set of data (x,y). We could randomly guess the m and b values of the line, tabulate the error of each, and then identify which guess results in the least error. The given equation can be estimated as

$$y = \overset{\frown}{m} x + \overset{\frown}{b}$$

This method is called a least squares linear regression method. We first minimize the error. The derivative of the error function reaches a minimum and its slope is equal to zero when the error value is also at a minimum. Let

$$\varepsilon^2 = error^2 = \Sigma \left(y_i - y_p \right)^2,$$

and,

$$y_p = mx_i + b + \varepsilon$$

where, y_p = predicted value when $x = x_i$, and y_i = experimental value when $x = x_i$.

Then as before,

$$\varepsilon^2 = \Sigma \left(y_i - m x_i - b \right)^2$$

$$\varepsilon^2 = \Sigma \left(m^2 x_i^2 + 2mbx_i - 2mx_i y_i + b^2 - 2by_i + y_i^2 \right) \quad (*)$$

As we approach the best value for m and b, ε^2 approaches its minimum value where the error as a function of m and b is at a minimum. Solve (*) for the minimum as a function of m. We seek the point where the first-derivative equals zero

$$\frac{d\varepsilon^2}{dm} = 0,$$

$$\frac{d\varepsilon^2}{dm} = 2m\Sigma x_i^2 + 2b\Sigma x_i - 2\Sigma \left(x_i y_i \right) = 0$$

$$2m\Sigma x_i^2 + 2b\Sigma x_i - 2\Sigma \left(x_i y_i \right) = 0$$

$$2b\Sigma x_i = 2\Sigma \left(x_i y_i \right) - 2m\Sigma x_i^2$$

$$b\Sigma x_i = \Sigma \left(x_i y_i \right) - m\Sigma x_i^2$$

$$b = \frac{\Sigma \left(x_i y_i \right)}{\Sigma x_i} - \frac{m\Sigma x_i^2}{\Sigma x_i}$$

$$b = \frac{\Sigma \left(x_i y_i \right) - m\Sigma x_i^2}{\Sigma x_i}$$

Solve (*) for the minimum as a function of b and seek the point where the first-derivative equals zero

$$\frac{d\varepsilon^2}{db} = 0$$

$$\frac{d\varepsilon^2}{db} = 2m\Sigma x_i + 2\Sigma b - 2\Sigma y_i = 0$$

$$2m\Sigma x_i + 2\Sigma b - 2\Sigma y_i = 0$$

$$2\Sigma b = 2\Sigma y_i - 2m\Sigma x_i$$

To simplify the notation (for number of points $= n$), let

$$S_x = \Sigma x_i, S_y = \Sigma y_i, S_{xy} = \Sigma(x_i y_i), S_{xx} = \Sigma(x_i^2), nb = \Sigma b$$

$$mx_i = y_p - b$$

From the above steps, solve for m:

$$m = S_{xy} - bS_x S_{xx}$$

Or, stated differently, $S_{xy} = mS_{xx} + bS_x$.

Box-Jenkins technique

The Box-Jenkins technique consists of a family of time-series models. It comprises of many different models. These models can be grouped into three basic classes – autoregressive (AR) models, MA models, and autoregressive integrated moving average (ARIMA) models. For many problems in business, engineering, and physical and environmental sciences, the Box-Jenkins technique may be applied to related variables of interest. An analysis may be obtained by considering individual series as components of a multivariate or vector time series and analyzing the series jointly. The Box-Jenkins technique is used to study the relationship among variables. This involves the development of statistical models and methods of analysis that describe the inter relationships among the series.

The models for time series are stochastic models. A time series $z_1, z_2, \ldots, z_n$ of N successive observations is regarded as a sample realization from an infinite population of such time series that could have been generated by the stochastic process. The *backward shift operator B* is defined by $Bz_t = z_{t-1}$; hence, $B^m z_t = z_{t-m}$. Another important operator is the *backward difference operator*, ∇, defined by $\nabla z_t = z_t - z_{t-1}$. This can be written in terms of B, since

$$\nabla z_t = z_t - z_{t-1} = (1-B)z_t$$

The stochastic models employed are based on the idea that an observable time series z_t in which successive values are highly dependent can frequently be regarded as generated from a series of independent "shocks" a_t. These *shocks* are random drawings from a fixed distribution, usually assumed normal and having mean zero and variance σ_a^2. Such a sequence of independent random variables $a_t, a_{t-1}, a_{t-2}\ldots$ is called a *white noise* process. The white noise process a_t is transformed into the process z_t by what is called a linear filter. The linear filtering operation simply takes a weighted sum of previous random shocks a_t, so that

$$z_t = \mu + a_t + \psi_1 a_{t-1} + \psi_2 a_{t-2} + \ldots$$

$$= \mu + \psi(B) a_t$$

In general, μ is a parameter that determines the "level" of the process, and

$$\psi(B) = 1 + \psi_1 B + \psi_2 B^2 + \ldots$$

is the linear operator that transforms a_t into z_t, and is called the transfer function of the filter. The model representation of equations given above can allow for a flexible range of pattern of dependence among values of the process $\{z_t\}$ expressed in terms of the independent random shocks a_t. The sequence $\psi_1, \psi_2, \ldots$ formed by the weights may, theoretically, be finite or infinite. If the sequence is finite, or infinite and absolutely summable in the sense that

$\sum_{j=0}^{\infty} \left| \psi_j \right| < \infty$, the filter is said to be stable and the process z_t is station-

ary. The parameter μ is then the mean about which the process varies. Otherwise, z_t is non-stationary and μ has no specific meaning except as a reference point for the level of the process.

Denote the values of a process at equally spaced times $t, t-1, t-2\ldots$ by $z_{t-1}, z_{t-2}, \ldots$.

Also let $\tilde{z}_t = z_t - \mu$ be the series of deviations from μ. Then

$$\tilde{z}_t = \phi_1 \tilde{z}_{t-1} + \phi_2 \tilde{z}_{t-2} + \ldots + \phi_p \tilde{z}_{t-p} + a_t$$

is called an *AR* process of order p. The reason for this name is that a linear model

$$\tilde{z} = \phi_1 \tilde{x}_1 + \phi_2 \tilde{x}_2 + \ldots + \phi_p \tilde{x}_p + a$$

relating a "dependent" variable z to a set of "independent" variables $x_1, x_2, \ldots x_p$, plus a random error term a, is referred to as a regression model, and z is said to be "regressed" on previous values of itself: hence the model is *AR*. If an *AR* *operator* of order p is defined in terms of the backward shift operator B by

$$\phi(B) = 1 - \phi_1 B - \phi_2 B^2 - \ldots - \phi_p B^p,$$

The AR model given may be written as

$$\phi(B)\tilde{z}_t = a_t$$

The model contains $p+2$ unknown parameters $\mu, \phi_1, \phi_2, \ldots \phi_p, \sigma_a^2$, which in practice have to be estimated from the data. The additional parameter σ_a^2 is the variance of the white noise process a_t.

The AR model is a special case of the linear filter model. For example, $\tilde{z}_{t-1}$ can be eliminated from the right-hand side of the AR model by substituting

$$\tilde{z}_{t-1} = \phi_1 \tilde{z}_{t-2} + \phi_2 \tilde{z}_{t-3} + \ldots + \phi_p \tilde{z}_{t-p-1} + a_{t-1}$$

Similarly, $\tilde{z}_{t-2}$ can be substituted, and so on, to yield eventually an infinite series in the a's. Consider, specifically, the simple first-order $(p=1)$ AR process, $\tilde{z}_t = \phi\tilde{z}_{t-1} + a_t$. After m successive substitutions of $\tilde{z}_{t-1} = \phi\tilde{z}_{t-1} + a_{t-j}, j=1,\ldots,m$, in the right-hand side we obtain

$$\tilde{z}_t = \phi^{m+1}\tilde{z}_{t-m-1} + a_t + \phi a_{t-1} + \phi^2 a_{t-2} + \ldots + \phi^m a_{t-m}$$

In the limit as $m \to \infty$ this leads to the convergent infinite series representation $\tilde{z}_t = \sum_{j=0}^{\infty} \phi^j a_{t-j}$ with $\psi_j = \phi^j, j \geq 1$, provided that $|\phi| < 1$. Symbolically, in the general AR case, we have that

$$\phi(B)\tilde{z}_t = a_t$$

is equivalent to

$$\tilde{z}_t = \phi^{-1}(B)a_t = \psi(B)a_t$$

with $\psi(B) = \phi^{-1}(B) = \sum_{j=0}^{\infty} \psi_j B^j$.

AR processes can be stationary or non-stationary. For the process to be stationary, the ϕ's must be such that the weights $\psi_1, \psi_2, \ldots$ in $\psi(B) = \phi^{-1}(B)$ form a convergent series. The necessary requirement for stationarity is that the AR operator, $\phi(B) = 1 - \phi_1 B - \phi_2 B^2 - \cdots - \phi_p B^p$, considered as a polynomial in B of degree p, must have all roots of $\phi(B)=0$ greater than 1 in absolute value; that is, all roots must lie outside the unit circle. For the first-order AR process $\tilde{z}_t = \phi\tilde{z}_{t-1} + a_t$ this condition reduces to the requirement that $|\phi| < 1$, as the argument above has already indicated.

The general form of the model that is used to describe time series is the ARIMA model

$$\varphi(B)z_t = \phi(B)\nabla^d z_t = \theta_0 + \theta(B)a_t$$

where:

$$\phi(B) = 1 - \phi_1 B - \phi_2 B^2 - \ldots - \phi_p B^p$$

$$\theta(B) = 1 - \theta_1 B - \theta_2 B^2 - \ldots - \theta_q B^q$$

$\phi(B)$ and $\theta(B)$ are polynomial operators in B of degrees p and q. This process is referred to as an ARIMA(p, q) process.

The ARIMA model can be expressed explicitly in terms of current and previous shocks. A linear model can be written as the output z_t from the linear filter

$$z_t = a_t + \psi_1 a_{t-1} + \psi_2 a_{t-2} + \ldots$$

$$= a_t + \sum_{j=1}^{\infty} \psi_j a_{t-j}$$

$$= \psi(B)a_t$$

whose input is a white noise, or a sequence of uncorrelated shocks a_t with mean 0 and common variance σ_a^2. Operating on both sides of the ARIMA model with the generalized AR operator $\varphi(B)$, then

$$\varphi(B)z_t = \varphi(B)\psi(B)a_t$$

However, since

$$\varphi(B)z_t = \theta(B)a_t$$

it follows that

$$\varphi(B)\psi(B) = \theta(B)$$

Therefore, the ψ weights may be obtained by equating coefficients of B in the expansion

$$\left(1 - \varphi_1 B - \ldots - \varphi_{P+d} B^{p+d}\right)\left(1 + \psi_1 B + \psi_2 B^2 + \ldots\right)$$

$$= \left(1 - \theta_1 B - \ldots - \theta_q B^q\right)$$

Thus, the ψ_j weights of the ARIMA process can be determined recursively through the equations

$$\psi_j = \varphi_1\psi_{j-1} + \varphi_2\psi_{j-2} + \ldots + \varphi_{p+d}\psi_{j-p-d} - \theta_j \quad j > 0$$

with $\psi_0 = 1, \psi_j = 0$ for $j < 0$, and $\theta_j = 0$ for $j > q$. It is noted that for j greater than the larger of $p+d-1$ and q, the ψ weights satisfy the homogenous difference equation defined by the generalized AR operator, that is,

$$\varphi(B)\psi_j = \phi(B)(1-B)^d \psi_j = 0$$

where B now operates on the subscript j. Thus, for sufficiently large j, the weights ψ_j are represented by a mixture of polynomials, damped exponential, and damped sinusoid in the argument j.

MAs

This MA is also referred to as smoothing technique. This method is suitable for forecasting data with no trend or seasonal pattern. The technique is a time series sequence of observations which are ordered in time. Inherent in the collection of data taken over time is some form of random variation. The technique, when properly applied, reveals more clearly the underlying trend, seasonal and cyclic components of a rime series.

The MA is the best-known forecasting method. It simply takes a certain number of past periods and adds them together; then divides by the number of periods. Simple MA is an effective and efficient approach provided the time series is stationary in both mean and variance. The following formula is used in finding the MA of order n, MA(n) for a period t+1,

$$MA_{t+1} = [D_t + D_{t-1} + \ldots + D_{t-n+1}] / n$$

Where D_i are past periods, and n is the number of observations used in the calculation. The forecast for time period t+1 is the forecast for all future time periods. However, this forecast is revised only when new data becomes available.

The weighted MA is another form of MAs. It is very powerful and economical. They are widely used where repeated forecasts required methods like sum-of-the-digits and trend adjustment methods. As an example, a weighted MA is:

$$Weighted\ MA\ (3) = w_1.D_t + w_2.D_{t-1} + w_3.D_{t-2}$$

where the weights are any positive numbers such that: w1+w2+w3=1. The average weights for this example are $w_1 = 3/(1+2+3) = 3/6$, $w_2 = 2/6$, and $w_3 = 1/6$.

Example

The MA and weighted MA of order five are calculated in Table 8.3. The computations are calculated using the weighted MA formula.

ANNs

ANNs are used to generate a mapping between some input data and some required output. ANNs are model free estimators in that they do not rely on an assumed form from the underlying data. Rather, based on some observed data, they attempt to obtain an approximation of the underlying system that generated the observed data. They use a non-linear data driven self-adaptive approach as opposed to the traditional model-based methods. They are powerful tools for modeling, especially when the underlying data relationships are known. ANNs can identify and learn the correlated patterns between the input data set and the corresponding target outputs. After training, the ANNs can be used to predict the outcome of new, unseen independent input data. The ANNs imitate some aspects of the structure and learning of the human brain and can process problems involving non-linear and complex data even when the data are imprecise and noisy. One of the appealing features of ANNs is that "learning by example" replaces "programming" in solving problems. This feature makes such computational models very useful in applications where one has little or incomplete understanding of the problem to be solved, but where training data is readily available. ANNs are not intelligent, but they are good at recognizing patterns and making simple rules for complex problems. The neural network was historically inspired by the biological functioning of a human brain. Specifically, it

Table 8.3 Computations for weighted MAs

Week	Sales ($1000)	MA(5)	WMA(5)
1	105	–	–
2	100	–	–
3	105	–	–
4	95	–	–
5	100	101	100
6	95	99	98
7	105	100	100
8	120	103	107
9	115	107	111
10	125	117	116
11	120	120	119
12	120	120	119

attempts to mimic the fault-tolerance and capacity to learn of biological neural systems by modeling the low-level structure of the brain. The brain is composed of a very large number of interconnected neurons. The neuron has a branching input structure (the dendrites), a cell body (the soma), and a branching output structure (the axon). The axon of one cell connects to the dendrites of another through a synapse (Figure 8.6). When a neuron is activated, it fires an electrochemical signal along the axon. The signal crosses the synapses to the other neurons, which may in turn fire. A neuron fires only if the total signal received at the cell body from the dendrites exceeds a certain level (the firing threshold). The chance of firing, i.e., the strength of the signal received by a neutron depends on the efficacy of the synapses.

When creating a functional model of the biological neutron, there are three basic components of importance.

1. *Synapses*: This is modeled as weights in ANNs. The strength of the connection between an input and a neuron is represented by the value of the weight. Unlike the synapses in the brain, the synapse weight of the artificial neuron lies within the range of positive and negative values. Positive weight values designate excitatory connections while negative values reflect inhibitory connections.

The next two components model the activity within the neuron cell:

2. *Linear combination*: This is an adder to sum up the input signal modified by their respective weights.
3. *Activation function*: This controls the amplitude of the output of the neuron. An acceptable range of output is usually between 0 and 1, or −1 and 1.

This model is described schematically in Figure 8.7.

The model illustrated in Figure 8.7 includes an externally applied bias, b_k, which has the effect of increasing or lowering the net input of

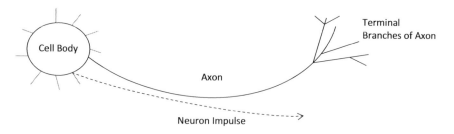

Figure 8.6 Structure of a neuron.

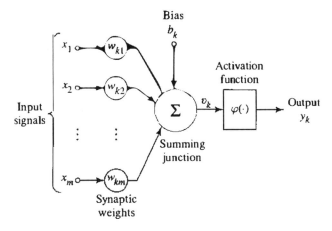

Figure 8.7 Artificial neuron model.

the activation function, depending on whether it is positive or negative, respectively. Mathematically, using Figure 8.7, a neuron k can be written as:

$$u_k = \sum_{j=1}^{m} w_j x_j$$

and

$$y_k = \varphi(u_k + b_k)$$

where $x_1, x_2, \ldots, x_m$ are the input signals; $w_1, w_2, \ldots, w_m$ are the respective synaptic weights of neuron k; u_k is the linear combination output due to the input signal; b_k is the bias; $\varphi(.)$ is the activation function and y_k is the output signal of the neuron. The use of bias b_k has the effect of applying an affine transformation to the output u_k:

$$v_k = u_k + b_k$$

where v_k is termed as the induced local field or activation field of neuron k. The linear combiner u_k is modified by b_k in the manner shown in Figure 8.8.

A bias is similar in function to a threshold and is treated as a weight connected to a node that is always on. The weights determine where this hyperplane lies in the input space. Without a bias term, this separating hyperplane is constrained to pass through the origin of the space defined by the inputs. For some problems this is acceptable, but in many problems the hyperplane would produce increased performance away from the

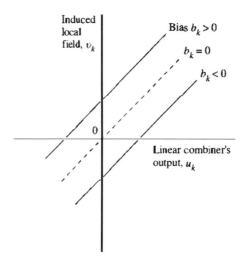

Figure 8.8 Affine transformation produced by the presence of a bias.

origin. If there are many inputs in a layer, they share the same input space and without bias they would all be constrained to pass through the origin. There are two fundamentally different classes of network architecture:

i *Feedforward ANNs*

In this type of topology, the connections between the neurons in an ANN flow from input to output only. These ANNs can be further divided into either single-layer feedforward ANNs or multi-layer feedforward ANNs. The single-layer network is the simplest form of a layer network that has only one input layer that links directly to the output layer. Figure 8.9(a) shows a one-layer network with the single layer referring to the output layer. With multi-layer feedforward ANNs, one or more hidden layers are present between the input and output layers, as shown in Figure 8.9(b), by adding one or more hidden layers, the network can extract higher-order statistics from its input and model more complex non-linear models.

ii. *Feedback/Recurrent ANNs*

In feedback or recurrent ANNs, there are connections from later layers back to earlier layers of neurons. There is at least one feedback loop in this type of network. Either the network's hidden neuron unit activation or the output values are fed back into the network as inputs. The internal states of the network allow this type of network to exhibit dynamic behavior when modeling the data's dependence on time or space. With

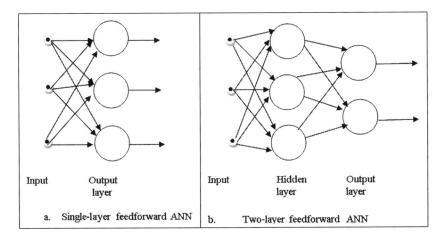

Input	Output layer	Input	Hidden layer	Output layer

a.   Single-layer feedforward ANN | b.   Two-layer feedforward ANN

Figure 8.9 Feedforward ANNs topology.

one or more feedback links whose state varies with time, the network has adjustable weights. This results in the state of its neuron being dependent not only on current input signal, but also on the previous states of the neuron. In other words, the network behavior is based on the current input and the results of previous processing inputs.

The learning process through which ANNs function can be categorized as supervised learning and unsupervised learning. For the purpose of clarity, the two classes of learning are explained as follows:

1. *Supervised learning*

This form of learning can be regarded as analogous to learning with a teacher, whereby the teacher has the knowledge of the environment. The knowledge is represented as input-output combinations. This environment is unknown to the neural network system. The teacher who has the knowledge of the environment will provide the neural network with desired responses for the training vectors. The network parameters are adjusted step by step under the combined influence of the training vector until the network emulates the teacher, producing the desired outputs for the corresponding inputs. At this stage, the network is presumed to be optimum in some statistical sense. In this way, the knowledge of the environment available to the teacher (ANN parameters frozen) is transferred to the neural network through training and stored in the form of fixed synaptic weights, representing long-term memory. When this condition is reached, the network is released from the teacher to deal with the environment by itself. With an adequate set of input-output examples, and enough time in which to do the training, a supervised learning system is usually able to approximate an unknown input-output mapping reasonably well.

2. *Unsupervised learning*

In this form of learning, the networks learn on their own as a type of self-study. When a set of data is presented to the network, it will learn to recognize patterns in the data. To perform unsupervised learning, a competitive-learning rule may be applied e.g., by creating a neural network with two layers – an input layer and a competitive layer. The input layer will receive the available data. The competitive layer consists of neurons that compete with others for the opportunity to respond to features contained in the input data. The output of the network is not compared with the desired output. Instead, the output vector is compared with the weight vectors leading to the competitive layer. The neuron with the weight vectors most closely matching the input vector is the winning neuron. In other words, the network operates in accordance with a "winner-takes-all" strategy. Once the network architecture is decided and the data needed are collected, the next phase of the ANN methodology is the training of the ANN model. The training goal is to find the training parameters that result in the best performance, as judged by the ANN's performance with unfamiliar data. This measures how well the ANN will generalize. To find the optimum ANN configuration, an ideal approach is to divide the data into three independent sets: training, validation, and testing. The definitions of these terminologies are given as follows:

> *Training set*: A set of examples used to adjust or train the weights in the ANN to produce the desired outcome.
> *Validation set*: The validation error is used to stop the training. The validation error is monitored to determine the optimum point to stop training. Normally, the validation error will decrease during the initial phase of training. However, when the ANN begins to overfit the data, the output error produced by the validation set will begin to rise. When the validation error increases for an appreciable number of iterations, thus indicating the trend is rising, the training is halted, and the weights that were generated at the minimum validation error are used in the ANN for the operation.
> *Testing set*: To assess the performance of the ANN. As the real prediction accuracy will generally be worse than that for the holdout example, there is a need to evaluate the developed model with some real problems. In this research, the set of independent data used will be termed "evaluation sets." The "evaluation sets" used in this research represent two real problems, i.e., predicting electricity consumption from two different data which were not used during the training process.

Big Data

For most organizations, Big Data is the reality of doing business. It is the proliferation of structured and unstructured data that flood organizations on a daily basis, and if managed well can provide powerful insights. Data is flowing and the volume is growing. With the massive generation of information from the advent of the internet and the increasing digitalization of business, there is tremendous opportunity in the new amounts and types of data collected. But this data explosion has also dramatically increased the complexity of the enterprise data landscape, with multiple data lakes, data warehouses, operational applications, e-commerce, online interactions, and so on.

Big Data technologies lack enterprise governance, holistic lifecycle management, and security concepts. These providers are just coming up the curve and trying to provide the level of enterprise governance and security that enterprise data warehouse (EDW) and database providers have been delivering for their offerings for years. That means that organizations are stuck with limited tools for integrating systems and creating data pipelines. As a result, it takes a lot of effort to create a data pipeline across the enterprise. Some companies have tried to solve these issues by maintaining two sets of data: one for transactions and one for analysis. But this is not only costly and inefficient; it also leads to discrepancies because it's hard to keep them in sync. And discrepancies lead to inaccurate analytical outcomes, with the obvious negative impact on decision-making.

Challenges of Big Data

Meeting the needs of business and the fast pace of today's demands means that enterprise data needs to overcome the following three challenges:

1. **Governance**
   We face the lack of visibility, and ask: Who changed the data? What was changed? Who is accessing it?

2. **Data pipeline**
   It is difficult to refine and enrich data across multiple systems. For example, this might involve improving the value of existing data by appending information, such as connecting sensor data with the asset ID and asset profile information, held in a different system.

3. **Data sharing**
   Unfortunately, integration is manual, point-to-point, painful, and slow. Changing an integration point usually depends on the agility and flexibility of the IT line.

A modern big data management strategy

The solution to better address these challenges would be a Big Data landscape and data operations management solution that enables agile data operations across the enterprise, and also enables data governance, pipelining, and sharing of all data in the connected landscape. The vision should be to provide the ability to understand, connect, and drive processes across the multiple data sources and endpoints with which the enterprise struggles today. By providing visibility into the landscape of data opportunities, as well as providing an easy way to connect data sources and create powerful data pipelines that hop across the landscape, businesses would be able to better achieve the data agility and business value that they seek.

Applying cloud enterprise resource planning (ERP) for businesses

ERP is a process by which a company (often a manufacturer) manages and integrates the important aspects of its business. An ERP management information system integrates areas such as planning, purchasing, inventory, sales, marketing, finance, and human resources. With cloud computing now considered mainstream, more and more organizations are embracing cloud ERP (public and private) to drive enterprise-wide innovation to improve customer relationships while delivering operational efficiencies.

This trend is set to continue, with analysts predicting that cloud ERP adoption globally will continue to accelerate in all markets for the foreseeable future. Cloud's ascendancy in the enterprise is also being driven by its business relevance. Where once cloud was just another emerging technology, it has now become the key contributor and enabler of many of the current technology megatrends including hyper-connectivity, Internet of Things, Big Data, and social media.

Despite the continuing growth of adoption rates, most organizations still lack a coherent cloud application strategy on which to execute. Many are jumping head first into the cloud with little consideration for its broader business value, with the primary focus on introducing new business capability or only replacing applications reaching end of life. This approach has the potential to undermine the true business value, as opportunities for tangible savings can be realized only once a critical mass of applications and data have been moved to the cloud. As such, a few key fundamental elements need to be clearly defined:

- A robust cloud strategy
- A supporting application and information strategy
- Establishment of a cloud-ready organization – people and process

These elements provide consistency in understanding the business drivers for cloud adoption, the architectural principles and framework relevant to the cloud, capabilities required both internal and external to the organization, approach to commission new and existing applications, and how risks associated with migration will be managed.

Building the Big Data warehouse

The EDW architecture has long been a key technology asset for fast analytics on cleansed, curated, and structured business data. It is a critical technology foundation for many enterprises. However, it is straining to deliver value in the era of exploding data volumes and increased demand for analytics and data across the organization. Not only are the sources of structured, traditional data increasing in volume – such as transactional, operational, and financial information – but organizations are also embracing the age of Big Data, dealing with new types of data that the EDW was not built to handle. This includes unstructured information, like weblog and machine log information, audio, video, and social media interactions; high-speed information, like sensor data in Internet-of-Things scenarios; and third-party information, like weather, public databases, or brokered information. All of this data is being introduced to the enterprise at unprecedented volume and speed. This Big Data is stored in systems uniquely capable of handling them, such as Hadoop-based data lakes or cloud object storage.

The business potential inherent in all of this data is demanding to be tapped, not only to improve the efficiency and quality of existing goods and services, but also to create new offerings or business models that can accelerate an organization ahead of the competition. However, in order to achieve this, enterprises need a way to interconnect Big Data with enterprise data. They also need a way to provide the analytical responsiveness, security, and ease of use that are associated with the EDW and its applications.

With a Big Data warehouse approach, companies are looking to:

- Leverage existing investments made in technology, processes, and people. While organizations recognize the limitations of the EDW in the face of new data demands, they also recognize the value of the data already being managed effectively by the EDW and want to leverage this as part of their new enterprise architecture. There is not only the financial investment in the technology itself to consider, but also the value of the existing processes that are well understood by the enterprise's employees, as well as partners and vendors. While change and new investment are inevitable, leveraging the EDW to its maximum potential makes clear sense from a financial and change management perspective.

- Innovate for the future, leveraging new and faster sources of data. Enterprises are increasingly embracing Big Data, especially as Big Data solutions become easier to use, more secure, and better integrated to traditional enterprise systems. These new, faster data sources represent greater opportunities for improving existing products and services, as well as an opportunity to capture new and emerging markets. Enterprises across industries increasingly see competitive threats from disruptive new entrants that are effectively using new data architectures built for Big Data from the start.

- Make faster, more responsive, and even proactive decisions. End customer expectations for service speed, corporate responsiveness, and information sharing from the companies that serve them are increasing quickly. As a result, enterprises are looking to deliver on expectations by accelerating their own speed of data collection, processing, and analysis, aiming to spread "right time" data and decision-making as broadly across their organizations as they can. Forward-looking organizations have future goals not only for responsiveness to rapidly changing market or customer conditions, but also for achieving predictive analytics that can help them to address issues before they escalate, or spot trending opportunities in their earliest stages.

- Empower more managers and decision-makers with analytics. "Self-service" analytics has been a goal of enterprises for some time now, and the intensity of pressure for achieving this goal has only increased. Organizations are realizing that limiting analytics and decision-making to an elite few requires too much time and can squander opportunities or exacerbate emerging challenges. By more broadly distributing information and analysis to a wider range of managers and decision-makers, including partners and key vendors, the organization becomes more responsive and agile.

- Ensure enterprise-class security and data governance, even as data volumes grow and analytical end users proliferate. While Big Data solutions have long offered a scalability and data diversity advantage that EDWs could not match, there have also been longstanding concerns about data security and data governance for these emerging technologies. As the technologies mature and can increasingly meet the security needs of the most demanding organizations, there is increased willingness to let Big Data projects out of the "lab" and interconnect them with the broader enterprise data architecture and its larger group of end users.

Big Data analytics

Big Data analytics refers to the strategy of analyzing large volumes of data, or Big Data. This Big Data is gathered from a wide variety of sources, including social networks, videos, digital images, sensors, and sales transaction records. The aim in analyzing all this data is to uncover patterns and connections that might otherwise be invisible, and that might provide valuable insights about the users who created it. Through this insight, businesses may be able to gain an edge over their rivals and make superior business decisions.

Big Data analytics allows data scientists and various other users to evaluate large volumes of transaction data and other data sources that traditional business systems would be unable to tackle. Traditional systems may fall short because they're unable to analyze as many data sources. Sophisticated software programs are used for Big Data analytics, but the unstructured data used in Big Data analytics may not be well suited to conventional data warehouses. Big Data's high processing requirements may also make traditional data warehousing a poor fit. As a result, newer, bigger data analytics environments and technologies have emerged, including Hadoop, MapReduce, and NoSQL databases. These technologies make up an open-source software framework that's used to process huge data sets over clustered systems.

Open-source Big Data analytics

Open-source Big Data analytics refers to the use of open-source software and tools for analyzing huge quantities of data in order to gather relevant and actionable information that an organization can use in order to further its business goals. Open-source Big Data analytics makes use of open-source software and tools in order to execute Big Data analytics by either using an entire software platform or various open-source tools for different tasks in the process of data analytics. Many Big Data analytics tools make use of open source, including robust database systems such as the open-source MongoDB, a sophisticated and scalable NoSQL database very suited for Big Data applications, as well as others. Open-source Big Data analytics services encompass:

- Data collection system
- Control center for administering and monitoring clusters
- Machine learning and data mining library
- Application coordination service
- Compute engine
- Execution framework

Data visualization

Data visualization is the presentation of data in a pictorial or graphical format. It enables decision-makers to see analytics presented visually, so they can grasp difficult concepts or identify new patterns. With interactive visualization, technology can be used to drill down into charts and graphs for more detail, interactively changing what data is seen and how it's processed. With Big Data there's potential for great opportunity, but many retail banks are challenged when it comes to finding value in their Big Data investment. For example, how can they use Big Data to improve customer relationships? How – and to what extent – should they invest in Big Data ?

Because of the way the human brain processes information, using charts or graphs to visualize large amounts of complex data is easier than poring over spreadsheets or reports. Data visualization is a quick, easy way to convey concepts in a universal manner – and you can experiment with different scenarios by making slight adjustments. Data visualization performs the following functions:

- Identify areas that need attention or improvement.
- Clarify which factors influence customer behavior.
- Helps understand which products to place where.
- Predict sales volumes.

Importance of data visualization

Data visualization has become the de facto standard for modern BI. Data visualization tools have been important in democratizing data and analytics and making data-driven insights available to workers throughout an organization. They are typically easier to operate than traditional statistical analysis software or earlier versions of BI software. This has led to a rise in lines of business implementing data visualization tools on their own, without support from IT.

Data visualization software also plays an important role in Big Data and advanced analytics projects. As businesses accumulated massive troves of data during the early years of the Big Data trend, they needed a way to quickly and easily get an overview of their data. Visualization is central to advanced analytics for similar reasons. When a data scientist is writing advanced predictive analytics or machine learning algorithms, it becomes important to visualize the outputs to monitor results and ensure that models are performing as intended. This is because visualizations of complex algorithms generally are easier to interpret than numerical outputs.

Usage of data visualization

Regardless of industry or size, all types of businesses are using data visualization to help make sense of their data. This is achieved as follows:

1. Comprehend Information Quickly

By using graphical representations of business information, businesses are able to see large amounts of data in clear, cohesive ways – and draw conclusions from that information. And since it's significantly faster to analyze information in graphical format (as opposed to analyzing information in spreadsheets), businesses can address problems or answer questions in a more timely manner.

2. Pinpoint Emerging Trends

Using data visualization to discover trends – both in the business and in the market – can give businesses an edge over the competition, and ultimately affect the bottom line. It's easy to spot outliers that affect product quality or customer churn, and address issues before they become bigger problems.

3. Identify Relationships and Patterns

Even extensive amounts of complicated data start to make sense when presented graphically; businesses can recognize parameters that are highly correlated. Some of the correlations will be obvious, but others won't. Identifying those relationships helps organizations focus on areas most likely to influence their most important goals.

4. Communication

Once a business has uncovered new insights from visual analytics, the next step is to communicate those insights to others. Using charts, graphs, or other visually impactful representations of data is important in this step because it's engaging and gets the message across quickly.

Conclusion

Over time, businesses have gathered a large volume of unorganized data that has the potential to yield valuable insights. However, this data is useless without proper analysis. Modeling techniques can find a relationship between different variables by uncovering patterns that were previously unnoticed. For example, analysis of data from point of sales systems and purchase accounts may highlight market patterns like increase in demand on certain days of the week or at certain times of the year. Optimal stock and personnel can be maintained before a spike in demand arises by acknowledging these insights. Modeling is not only great for lending empirical support to management decisions, but also for identifying errors in judgment. Modeling can provide quantitative support for decisions and prevent mistakes due to manager's intuitions.

Bibliography

Box, G., Jenkins, G. & Reinsel, G. (2008). *Time Series Analysis: Forecasting and Control.* 4th Edn. John Wiley Publishers, New York, pp. 100–105.

Craven, B. & Islam, S. (2005). *Optimization in Economics and Finance.* Springer Publishers, Dordrecht.

Hossein, A. (2017). *Time-Critical Decision Making for Business Administration.* University of Baltimore, MD, US.

Jarrett, J. (1991). *Business Forecasting Methods.* 2nd Edn. Cambridge Press, UK, pp. 22–34.

Jha, G. (2007). Artificial neural networks and its applications. Retrieved from www.ijeert.org/pdf/v2-i2/24.pdf 1 June 2016.

Kandananond, K. (2011). Forecasting electricity demand in Thailand with an artificial neural network approach. *Energies*, 4(12), 1246–1257.

Kaplan, R. & Norton, D. (1992). The balanced scorecard: Measures that drive performance. *Harvard Business Review*, 71.

Nissen, S. (2007). *Implementation of a Fast Artificial Neural Network Library (FANN).* Department of Computer Science University of Copenhagen (DIKU), Copenhagen.

Ozoh, P. (2016). Achieving Efficiency in Electricity Consumption Using Improved Machine Tools. Universiti Malaysia Sarawak. PhD Thesis.

Patterson, D. (1996). *Artificial Neural Networks.* Prentice Hall, Singapore.

Reichman, O., Jones, M. & Schildhauer, M. (2011). Challenges and opportunities of open data in ecology. *Science.* 331(6018), 703–705.

Segaran, T. & Hammerbacher, J. (2009). *Beautiful Data. The Stories Behind Elegant Data Solutions.* O'Reilly Media, p. 257.

Soo, S.C. (2010). An artificial neural network approach for soil moisture retrieval using passive microwave data. PhD Thesis. Curtin University of Technology, pp. 38–49.

Timm, L., Gomes, E., Barbosa, K., Reichardt, M., Souza, D. & Dynia, F. (2006). Neural network and state-space models for studying relationships among soil properties. *Scienticola Agricola*, 63, 386–395.

Tofallis, C. (2015). A better measure of relative prediction accuracy for model selection. *Journal of the Operational Research Society*, 66, 8, 1352–1362.

Yar, M. & Chatfield, C. (1990). Prediction intervals for the holt-winters forecasting procedure. *International Journal of Forecasting*, 6, 127–137.

Zhang, G.P., Areekul, P., Member, S., Senjyu, T., Member, S. & Toyama, H. (2010). A hybrid ARIMA and neural network model for short-term price forecasting in deregulated market. *IEEE Transactions on Power Systems*, 25(1), 524–530.

chapter nine

Integration of manufacturing and business models

Introduction

With the globalization and volatility of the markets, there is a huge pressure for manufacturing companies to be more innovative and competitive in delivering value to their customers. The evaluation of the overall value chain, designed and implemented to supply a specific product or service, should support changes in the existing business model or in the definition of new business models that ensures higher levels of customer satisfaction. The business model is at the core of the competitive response of any company to the market, defining the value proposition, the required activities, resources and partners, and knowledge of customers, costs, or profits related with its overall operation. A business model is a plan for the successful operations of a business, identifying sources of revenue, the intended customer base, products, and details of financing. Business models are used for a broad range of informal and formal descriptions to represent core aspects of a business, including purpose, business processes, targeting of customers, company offerings, strategies, infrastructure, organizational structures, sourcing, trading practices, operational practices, and policies. The relevance that integrated products and services is assuming nowadays, allowing manufacturing companies to achieve a longer and more stable relationship with their customers, determines new approaches to product-service development and methodologies, or tools to support the review and definition of appropriate business models.

To compete in terms of added value, manufacturing companies must be able to design and sustain their businesses based on a network of complementary capabilities to respond to market opportunities. The ability to design products or services and competitively delivery them into the market requires the definition, implementation, and management of new business models. Business models are often framed in response to particular competitive circumstances and it outlines how a company generates revenues with reference to the structure of its value chain and its interaction with their supplier, customers, and other partners with complementary competencies.

The relevance of business models in manufacturing companies

The global and sustainable competitiveness of manufacturing companies has a major impact in supporting economic growth and employment creation through delivery, added value products, and/or services. Therefore, manufacturing companies are each time more challenged to compete in terms of added value to achieve dominance in markets, since purely cost-based competition is not compatible with the goal of maintaining social and sustainability values. The added value is related to a company's ability to deliver customer-focused solutions, for example, adding services or integrating services into their core products. This trend, the servitization of manufacturing, is gaining more importance in our global economy. To support this trend, companies need methodologies to drive them in a paradigm shift that goes from considering products and services independently to considering them integrated.

The understanding of the product-service system (PSS) concept allows companies to shift their business focus in designing and delivering products to delivering a system of products and services. It requires the development of new relationships with stakeholders. The company becomes responsible for maintaining the product along its life cycle. A whole life-cycle business model is required and imperative. With PSS, customers have more customized offers, higher quality, and product-services with new functionalities. On the other hand, companies have new market opportunities, gain a new competitive advantage, and improve the total value delivered for the customer. The definition and implementation of new business models will enable the growth of new businesses and allow existing industries to sustain their global competitiveness. Business models need to be evaluated and managed by the manufacturing companies, adapting its characteristics to take advantage of the market conditions. Thus, the evaluation and redesign of business models will allow manufacturing companies to maintain their sustainability, promoting a more long-term relationship with their customers, innovating and supplying additional added value related to their products.

Industries have defined "frameworks" for business modeling. These frameworks attempt to define a rigorous approach to business value streams. Business model frameworks represent the core aspect of any industry; they involve the totality of how an industry selects its customers, configures its resource, goes to market, creates utility for customers, and captures profits. Through a constant monitoring and evaluation of the business model, industries can be at the forefront of their business market. The best approach to clarifying the process is to map the business model, which allows companies to experiment with different alternatives. The business model can be divided into nine components: (1) customer segments, (2) value proposition, (3) customer relationships, (4) channels, (5) key resources, (6) key partners, (7) key activities, (8) cost structure, and (9) revenue streams. See Figure 9.1.

Customer segments

Customer segments are the community of customers or businesses that a company is aiming to sell a product or service to. This is one of the most important building blocks in the business model for a business, so getting this building block right is the key to success. Customers can be segmented into distinct groups based on needs, behaviors, and other traits that they share. A customer segment may also be defined through demographics such as age, ethnicity, profession, gender, etc. or other psychographic factors such as spending behavior, interests, and motivations. An organization can choose to target a single group or multiple groups through its products and services.

By matching customer segments to a company's value proposition, a company can achieve a more lucrative revenue stream. An organization can categorize customers into distinct groups, if they have the following characteristics:

1. The customer groups have a particular need which justifies the creation of a product to match this need.
2. The group needs a separate distribution channel to be reached.
3. The groups require relationships of different kinds.
4. There is a very clear difference in the level of profitability each group represents to the organization.
5. Each consumer group feels strongly enough to pay for a different version of the product or service, tailored to their preferences.

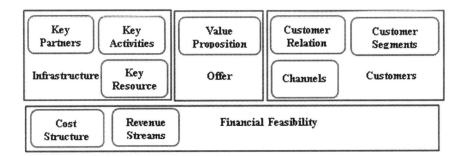

Figure 9.1 Business model components.

Value proposition

A value proposition makes a case for why a customer should pick one product over another, citing the unique value the product provides for its customers. It provides a unique combination of products and services to the customer by resulting in the solution to a problem the customer is facing or in providing value to the customer. This is the point of intersection between the product and the reason behind the customer's impulse to buy. A product can have a single or multiple value propositions.

Customer relations

Customer relations are the building blocks that describe the types of relationships a company establishes with specific customer segments. Customer relationships may be driven by one or more of three motivations:

1. Customer acquisition
2. Customer retention
3. Increased sales

   Industries need to be clear about their motivations. They need to analyze performance carefully to establish such benchmarks as: cost of customer acquisition, effectiveness of various marketing approaches, average period of customer retention, and integration of motivations with overall company policy.

Channels

Channels describe how a company communicates with and reaches its customer segments to deliver its value proposition. It is important to understand which pathway (or channel) is best for a company to reach its customers. Channel functions include the following:

1. Raising awareness among customers about a company's products and services
2. Helping customers evaluate a company's value proposition
3. Allowing customers to purchase specific products and services
4. Delivering a value proposition to customers
5. Providing post-purchase customer support

There are five types of channel phases. They are:

1. *Awareness* – How is awareness raised about the company's products and services?
   a. Advertising (word of mouth, social media, newspaper, etc.)

2. *Evaluation* – How are customers helped to evaluate an organization's value proposition?
   a. Surveys
   b. Reviews
3. *Purchase* – How are customers helped to purchase specific products and services?
   a. Web vs. brick and mortar
   b. Self checkout
4. *Delivery* – How is value proposition delivered to customers?
   a. Over the counter
   b. Delivered/Catered
5. *After Sales* – How is post-purchase customer support delivered?
   a. Call center
   b. Return policy
   c. Customer assistance

Holistic approach to optimizing a business's value chain

Manufacturing companies can start the digitalization process at any point in their value chains. The integration of manufacturing and business models to enterprise solutions enables manufacturing companies to integrate and digitalize their business processes – including their suppliers. They can start at any point in their value chain, from product design to production planning, production engineering, production execution, and services, and expand the digitalization process step by step. This process goes from machine design to engineering, commissioning, machine operation, and services. This means manufacturers can analyze their production facilities and products in actual use and feedback the insights into the entire value chain for continuous optimization.

The development of new business models

The development of an innovative product-service strategy is a clear response to the global market and requires adapting the business models to new challenges and the increase of service content. This strategy is based on the development of solutions for customers, strongly supported by core competencies on systems integration involving the provision of services, rather than just design, development, and distribution of high-quality products. This implies, of course, changes in the planning, conceptualizing, and implementing of integrated PSSs. Manufacturing companies need to be supported in the definition, implementation, and management of new business models.

There is a sequence of phases and activities that support the conception of new business models, its implementation plan, and management. This is an iterative process consisting of several loops. A good roadmap should be simple, logical and intuitive, comprehensive, and easy to implement. The roadmap consists of four phases: analysis, design, implementation, and evaluation. Each phase is divided into a set of activities (see Figure 9.2). The phases are described as follows:

Analysis phase

The analysis phase focuses on the characterization of a company's internal and external environment. A good understanding of the contextual environment contributes to the development of a competitive and efficient strategy. On the internal viewpoint, the aim is to look for the main weaknesses and strengths. On the external viewpoint, the goal is to look for opportunities and threats. This phase can be divided into four activities: needs characterization, internal analysis, external analysis, and requirements definition. The scope of the needs characterization activity is to formalize the need for a new business model that goes through the market and competition characterization, as well as a study of the existent business models. In the internal analysis, the aim is to analyze the strengths and the weaknesses of the company and minimize organizational and technological potential problems. In the external analysis activity, the contextual environment is examined and the opportunists and threats that surround the market environment are characterized. To do this, a set of tools such as Porter analysis, benchmarking, and a political, economic, social, and technological analysis are applied (see Figure 9.3). Finally, in

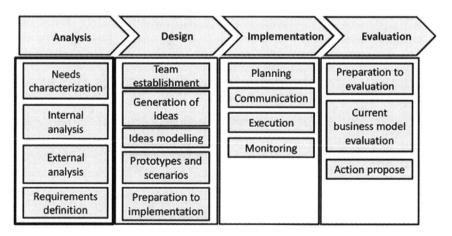

Figure 9.2 Business model roadmap with each phase divided into a set of activities.

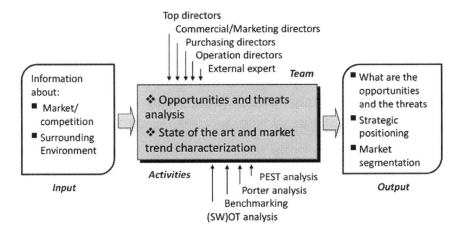

Figure 9.3 External analysis activity. Description of the input, the involved team, the main activities, and outputs.

the requirements definition the main goal is to realize the aims and goals of the business model.

Design phase

The design phase consists of the creation and development of new business models. This is a creative and innovative phase. The design phase is divided into six activities: team establishment, generation of ideas, ideas modeling, prototypes and scenarios, and preparation to implement. In the team establishment activity, the aim is to define the team members profile and choice of the members. The team should be multidisciplinary. The commitments must be defined as well. In ideas generation, the scope is to do some brainstorming and together come up with a set of ideas to the business model. This activity is followed by the modeling and selection of the best proposal that will converge in the business model prototype. These activities are shown in Figure 9.4.

Implementation phase

The implementation phase includes operational activities, such as the definition of work plans and schedules. This phase is divided into four activities; planning, communication, execution, and monitoring. The planning activity consists of essentials in elaborating a business plan, an activities plan, and the identification of milestones. Then, the business model, the business plan, and the plan of activities must be presented and explained to all partners. This is fundamental to all involved in the team. After the

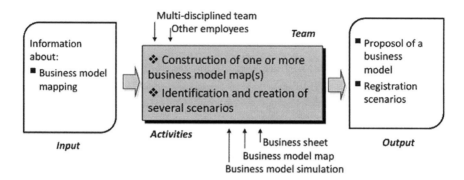

Figure 9.4 Prototypes and scenarios activity. Description of the input, the involved team, the main activities, and output.

communication, the business model is ready for execution. A constant follow-up must be done and all constraints should be reported. This continuous monitoring will enable going back to the roadmap for correction or stoppage of work if required.

Evaluation phase

The aim of the evaluation phase is to compare results that were obtained with the expected performance and draw conclusions with the intention to make changes to the current business model. This phase is divided into three activities: preparation to evaluation, current business model evaluation, and action propose. The preparation to evaluation consists of defining the key performance indicators and performing a survey to evaluate stakeholder's satisfaction. The evaluation of the business model properly consists of a critical reflection of the contextual environment using, for instance, the Strengths, Weaknesses, Opportunities, and Threats (SWOT) analysis (see Figure 9.5). The last activity is the action proposal. The output of the previous activity must be evaluated and changes must be proposed, communicated, and applied.

Solution to integrated business models

Most system integrators normally deliver custom solutions to solve particular problems of single customers. Some system integrators have also worked to create "productized solutions" that aim to be sold to a set of customers and that can be configured by the user and/or by self-learning. The integration of manufacturing and business models will look more like productized solutions as they allow scalability – a key element. Custom and productized solutions place very different requirements on

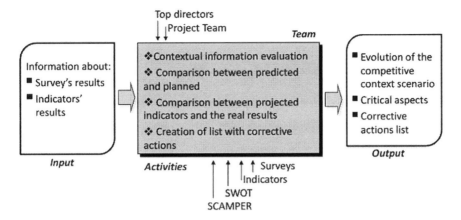

Figure 9.5 Current business model evaluation activity. Description of the input, the involved team, the main activities, and the outputs.

the system integrator's organization. The productized solution most likely requires:

- A solid understanding of a particular market need for which the solutions are developed.
- An ability to extrapolate the understanding to a broader set of potential customers.
- A proactive selling approach. An important set of system integrators don't manage their sales through dedicated sales resources but base it on repeat business, word of mouth, or proactive sales efforts by the owner.
- An ability to prototype quickly and also the ability to fail, recognize the failure quickly, learn from the mistakes, and move on.

Ideation processes are not very common among system integrators, but will become necessary if they want to evolve into the productized solutions space. By working closely with their customers, system integrators can develop an understanding of underserved needs. Developing some internal practices could help them make ideation an ongoing process. There are five steps to better business results. They are given as:

Step 1: Sense

The Sense step is the process of building the demand plan at the local and global company level. You'll seek to understand, or sense, how your plants and supply chain will need to respond to demand for your products and services. Sensing demand is a key step in avoiding both inventory shortages and overstocked items. Once you've developed a demand plan, you'll want to be sure that the plan is being fulfilled at least at the departmental level of your organization. Your demand planners or analysts will then review your forecast against customer service levels and inventory levels at the beginning of each month. This will help your decision-makers sense challenges from more angles so that you can update your forecast accordingly.

Step 2: Shape

The Shape step is the process of building a forward-looking aggregated and optimized demand and supply plan at the local and global level. In this step, you'll optimize demand with supply, identifying gaps and analyzing alternatives for addressing mismatches – before they affect the customer. You'll focus on building a strategic demand plan at the family or sub-family level of your product line, aiming for at least 18–24 months of future visibility. By the end of the Shape step, your optimized demand and supply plan will be complete.

Step 3: Collaborate

During the Collaborate step, you'll build strong alignment between your local, regional, and global teams, including the customer and suppliers. Your demand plans and supply forecasts can both be improved through active collaboration with your key stakeholders – across your enterprise and beyond. This collaboration should include the sales organization, trading partners, suppliers, and customers.

Technology is the collaboration enabler. Many leading firms are already using social media to enhance collaboration – and the trend will only ramp up in the future. In the Collaborate phase, you'll establish departmental coordination as well as ongoing collaboration outside the four walls of your company.

Step 4: Integrate

The Integrate step is the process of bringing together all executives and stakeholders to approve the optimized demand, supply, and financial plans. Don't forget that: Integrated Business Planning (IBP) requires buy-in across the organization as well as the extended supply chain. At this point, there's a single view of the truth. Everyone in the extended supply chain is operating with the same numbers and the same assumptions.

Don't underestimate the challenge of integration. While it's true that automated work-flow and other technologies can help, integration requires process discipline. At the end of the Integrate step, you'll end up with a single, optimized plan – and your financial plans will factor into all your supply chain management decisions.

Step 5: Orchestrate

The Orchestrate step is the process of publishing the integrated plan and adapting to the changes in the business, thus allowing for continuous improvement. It's here we see that building the plan isn't enough. Organizations are systems of people and interrelated functions and processes. You must operationalize and disseminate the IBP across your extended supply chain to optimize the flow of materials in ways that fulfill demand.

Leveraging a single IBP platform can dramatically simplify your orchestration process. By the end of the Orchestrate step, you will have achieved global cooperation with all your trading partners. You'll have a responsive supply chain that meets sudden and short-term fluctuations in demand.

Conclusion

The integration of manufacturing and business models to the enterprise is the activity of representing processes of an enterprise, so that the current process may be analyzed and improved in future. This is typically performed by business analysts and managers who are seeking to improve process efficiency and quality. The process improvements identified by business process modeling may or may not require information technology involvement, although that is a common driver for the need to model a business process by creating a process master.

The modeling of the enterprise and its environment could facilitate an enhanced understanding of the business domain and processes of the extended enterprise, and especially those that hold the enterprise together and extend across the boundaries of the enterprise. Thus, a fast understanding can be achieved throughout the enterprise about how business functions are working and how they depend upon other functions in the organization.

Bibliography

Baden_Fuller, C. and Mangematin, V.(2013). Business models: A challenging agenda, *Strategic Organization*, 11(4), 418–427.

Barquet A, Cunha V, Oliveira M and Rozenfeld, H. (2011). Business model elements for product-service system. *Proceedings of the 3rd CIRP International Conference on Industrial Product Service Systems*, Germany.

Berglund, H. and Sandstrom, C. (2013). Business model innovation from an open systems perspective: Structural challenges and managerial solutions, *International Journal of Product Development*, 3(4), 274–284.

Debei, A., El-Haddadeh, M. and Avison, D. (2008). Defining the business model in the new world of digital business. *Proceedings of the 14th Americas Conference on Information Systems*, Toronto, 1–11.

George, G. and Bock, A. (2011). The business model in practice and its implications for entrepreneurship research, *Entrepreneurship Theory and Practice*, 35(1), 83–111.

Hummel, E., Slowinski, G. and Gilmont, E. (2010). Business models for collaborative research, *Research Technology Management*, 53(6), 51–54.

Lindon, D., Lendrevic, J., Lévy, J., Dionísio, P., and Rodrigues, V. (2010). *Mercator XXI Teoria e Prática do Marketing*. 13ᵃ ed. Alfragide: D. Quixote.

chapter ten

Design of experiment techniques

Introduction

This chapter presents how two-level factorial experiments can be used to study how a response variable is influenced by certain factors for quality and process improvement. It will also be used to assess the effects of changing one factor independent of other factors. Topics covered in this chapter include the construction of factorial designs, the use of Pareto charts as a tool for estimating the effects of factors, the analysis of factor effects using ANOVA (analysis of variance) as well as the use of linear response surface methodology for optimizing the key quality characteristics of the process. In this chapter, we demonstrated that the two-level factorial design with interactions is more superior than the one-variable-at-a-time experimentation commonly used by engineers and scientists.

Factorial designs

A factorial experiment is an experiment designed to study the effects of two or more factors each of which is applied at two or more levels. In a balanced classical factorial experiment, all combinations of all the levels of the factors are tested. In this chapter, we consider only two-level factorial designs. In a 2k factorial design, 2 represents the number of levels and k represents the number of factors.

A factorial experiment can be used to study how a response variable is influenced by certain factors. It can be used to assess the effect of changing one factor independent of other factors. The premise of factorial experiments is that an observed response may be due to a multitude of factors. Since a dependent variable interacts with its environment, it is important to assess the simultaneous effects of more than one factor on the dependent variable. The example below shows a case where one response variable is influenced by two independent factors. Each factor is to be studied at three different levels.

Response variable: epoxy strength

Factor 1: Temperature (75° F, 80° F, 85° F)
Factor 2: Chemical concentration (high, medium, low)

Advantages of a factorial experiment

A factorial experiment has several advantages including those presented below:

Efficiency

- More robust compared to traditional single-factor experiments.
- In the one-factor study, it may be difficult to identify which one factor should be studied.
- More flexibility.

Information content

- More information can be derived from factorial experiments compared to single-factor experiments.

Validity of results

- Inclusion of multiple factors increases the validity of results.
- Results can identify direction for further experiments.

Factor

This is an independent variable or condition that is likely to affect the response or quality characteristic of interest. A factor may be a continuous variable such as oven temperature, RPM, pump pressure, webspeed, etc., or may be discrete (qualitative) variable such as catalyst type (A or B), valve (on or off), material type (A or B), cooling step (wet or dry), etc. Temperature, pressure, RPM, etc., are factors that can be controlled and measured. Therefore, they can be regarded as controllable and measurable factors. However, factors such as percent moisture going into an oven, and ambient humidity are measurable but uncontrollable. These factors are known as covariates. Other factors which are uncontrollable and immeasurable are useful in defining experimental error.

Levels

These are the settings of various factors in a factorial experiment such as high and low values of temperature, pressure, etc. For example, if the range of temperature to be studied is between 120° F and 180° F, then the low level can be set at 120° F and a high-level set at 180° F.

Response

Response is the measurement obtained when an experiment is run at each level of the factors under study. Responses may be continuous (quantitative) variables such as adhesion, percent yield, smoothness, etc., or discrete (qualitative) variables such as good or bad tastes, corrosion or no corrosion, etc. Rating scales can be used for qualitative variables as can be seen later in this chapter.

The basic layout of a factorial design is presented in Figure 10.1 for a two-factor experiment. Factor A has a levels. Factor B has b levels. There are n replicates for each cell. Each cell in the layout is referred to as a treatment which represents a specific combination of factor levels.

There are a total of $N = abn$ observations in the layout. Factorial designs are referred to as 2^f, 3^f, and so on. In a 2^f design, there are f factors, each having two levels. In a 3^f design, there are f factors, each having three levels.

There are three possible models for a factorial experiment depending on how the factor levels are chosen.

Fixed model

In this model, all the levels of the factors in the experiment are fixed.

Factor B Levels	Factor A Levels (sums, averages, etc.)				Row Summary
	$i=1$	$i=2$	....	$i=a$	
$j=1$	y_{111} y_{112} . . y_{11n}				α_1 α_2 . . .
$j=2$			$\bar{y}_{ijk}$		.
$j=3$					.
. . $j=b$					. . α_b
Column Summary	β_1	β_2	....	β_a	....

Figure 10.1 Layout of data collection for a two-factor factorial design.

Random model

In this model, the levels of the factors in the experiment are chosen at random.

Mixed model

In this model, the levels of some of the factors in the experiment are fixed while the levels of some of the factors are fixed.

The statistical model for the factorial experiment in Figure 10.1 is presented below:

$$Y_{ijk} = \mu + A_i + B_j + (AB)_{ij} + \varepsilon_{k(ij)}$$

where:

$i = 1, 2, \ldots, a$
$j = 1, 2, \ldots, b$
$k = 1, 2, \ldots, n$
$A_i = $ effect of the ith level of factor A
$B_j = $ effect of the jth level of factor B
$(AB)_{ij} = $ effect of the interaction between A_i and B_j
$Y_{ijk} = $ observation for the kth replicate of the A_i and B_j combination
$\varepsilon_{k(ij)} = $ random error associated with each unique combination of A_i and B_j.
$\mu = $ population mean.

The error terms are assumed to be independent and identically distributed normal variates with a mean of zero and variance σ_ε^2.

Experimental run

A run is when each control factor is set or fixed at a specific level and the experiment is run at those levels for the factors under study. For example, if an experimenter selects a pressure of 20 psi, a temperature of 150° F and a valve that is open, then this combination will represent a run.

One-variable-at-a-time experimentation

A one-variable-at-a-time experiment can be demonstrated by considering three factors A, B, and C. Let us say initially, each factor is set at their low levels. This means that A is at low, B is at low, and C is at low. When this level is run, the response y is 45. Now to determine the effect of factor A, the level of A is changed from low to high, and factors B and C still remain at low. Under this condition, the response value y is 20.

The change in the response value from 45 when all the three factors were set at their low levels, to 20 when only factor A was changed from low to its high level, can only be due to the effect of factor A and or experimental error. Similarly, the investigator can determine the effect of factor B by setting A at low, B at high, and C still remain at low, and obtains a response value at this new level, say 39. The investigator can now compare this 39–45 obtained for the control (run #1) to determine the effect of factor B. The setup can be described as shown in Figure 10.2. This is what is known as a one-variable-at-a-time experiment.

When experimenting with more than one factor, the one-variable-at-a-time experimental approach is inefficient and can provide misleading results. This type of experimentation has several serious problems. Some of these problems are:

- The effect of each factor is known at only one chosen level of each of the other factors.
- The effect of each factor is separated in time from the effect of other factors. Unknown extraneous factors which vary with time may therefore influence or bias the real effect of any factor under study.

Let us now consider a larger picture of a one-variable-at-a-time experiment. The objective of this experiment is to minimize MAG, an undesirable tar-like by-product. The investigator considered two variables, temperature and concentration, and studied the effects of these two variables on the response MAG. By setting all factors constant including the temperature and allowing the concentration to vary between 10% and 90%, the result (Figure 10.3) shows that MAG is minimum at about 28% concentration.

Then the investigator held the concentration constant at 28% and held all other variables constant as well, but varied temperature between 20 and 120° F. The result (Figure 10.4) shows that MAG is minimized at a temperature of 76° F. From this result one can conclude that the minimum MAG we can obtain is 20 at a temperature of 76° F and concentration of 28%.

	Factors				Response
Run	A	B	C		Y
1	−1	−1	−1	(control)	45
2	1	−1	−1		20
3	−1	1	−1		39
4	−1	−1	1		52

Figure 10.2 Design setup for one-variable-at-a-time experiment.

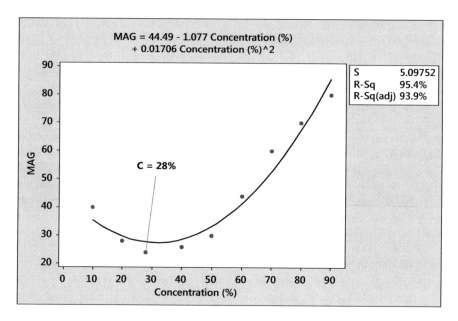

Figure 10.3 Effect of concentration on MAG.

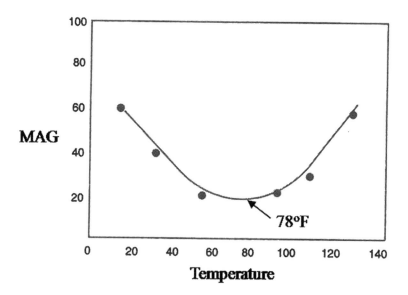

Figure 10.4 Effect of temperature on MAG.

Figure 10.5 shows a contour plot of temperature and concentration with MAG values plotted within the experimental region. As can be seen, a minimum MAG value of 5 can be achieved at a region of 80% of concentration and 40° F temperature. This example, therefore, demonstrates how a one-variable-at-a-time approach can fail to estimate the effects between two or more factors because the effect of temperature depends on the levels of concentration in this example. This effect between factors is known as interaction, and a one-variable-at-a-time experimental approach lacks the capability of detecting it.

The factorial experiment is superior to the one-variable-at-a-time experiment because of the following reasons:

- It allows the study of the effects of several factors in the same set of experiment.
- It provides the ability to test for the effect of each factor at all levels of the other factors and determine if this effect changes as the other factors change.
- It is capable of providing not only estimates of the effects separately (main effects) but also the joint effects of two or more factors (interaction effects).
- It provides a complete picture of what is happening over the entire experimental region than the one-variable-at-a-time.

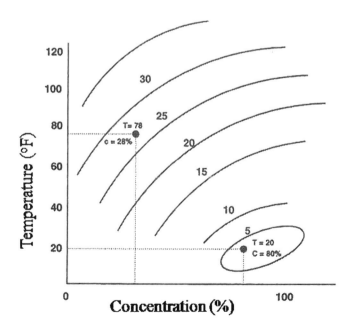

Figure 10.5 Contour plot of MAG with respect to temperature and concentration.

A good factorial experiment should incorporate basic experimental concepts such as randomization, replication, and orthogonality, as well as the iterative nature of experimentation such as conjecture, design, and analysis.

Randomization

Randomization in a design means running the order of experiment in a random (non-systematic) fashion. This eliminates or balances out the effects of undesirable systematic variation.

Replication

Replication is the running of the same set of conditions more than once. It is very important that for a true replication to occur and be distinguished from duplication, one should run the actual set of the condition to be replicated first, record the response, then change at least one or more of the levels, run the experiment at the new levels, and record the response; then come back and run the actual set of the replication again. By running replicate conditions back to back, one would be unable to account for variations that occur due to changes in raw material, operators, etc. In the analysis of the experimental results, it is also important to have an estimate of experimental error (random error) so as to have a meaningful yardstick for determining if estimated effects are real or due to common causes of variability only. Replication runs can be used to provide the estimate of the experimental error.

Orthogonality

Orthogonality in a design implies that the estimates of the main effects and interactions are uncorrelated with each other. Designs having this property ensure that if a systematic change occurs corresponding to any one of the effects, the change will be associated with that effect alone.

Experimenting with two factors: 2^2 factorial design

Conjecture

A process engineer wants to investigate the effects of two elements, nickel and gold, on the ductility of a new product. The ranges for these variables are as follows:

	Nickel (%)	Gold (%)
Low (−)	10	5
High (+)	20	10

The hypotheses that we will be testing are:

H_o = Effects are equal to zero (No effects exist)
H_A = Effects are not equal to zero

Design

The design matrix in standard order together with response values for a 2^2 design are presented in Table 10.1.

The design point should not be confused with run order. The design point should always be randomized to obtain the run order. The geometry of the design is presented in Figure 10.6.

Table 10.1 Design matrix for 2x2 study of ductility

Design point	Coded units			Un-coded units		
	Nickel	Gold	Strength	Nickel	Gold	Strength
1	−1	−1	52	10	5	52
2	1	−1	58	20	5	58
3	−1	1	75	10	10	75
4	1	1	64	20	10	64

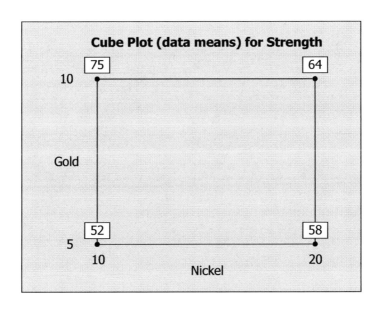

Figure 10.6 Design points geometry for ductility study.

Analysis

Estimation of effects: To calculate an effect, we use the equations below:

$$\hat{Y} = \text{response average}$$

$$\text{Effect Estimate} = \hat{Y}_{High} - \hat{Y}_{Low}$$

I. To estimate the effect of nickel:

$$\text{Effect of nickel} = \frac{58 + 64}{2} - \frac{52 + 75}{2}$$

$$= \hat{Y}_H - \hat{Y}_L$$

$$= 61.0 - 63.5$$

$$= -2.5$$

II. To estimate the effect of gold:

$$\text{Effect of nickel} = \frac{75 + 64}{2} - \frac{52 + 58}{2}$$

$$= 69.5 - 55$$

$$= 14.5$$

Interpretation

When the amount of nickel is changed from 10% nickel to 20% nickel, the effect on average is a reduction of 2.5 units on the breaking strength, while changing the amount of gold from 5% to 10% increases the breaking strength on average by 14.5 units.

Interaction

The interaction effect is the extent to which the effect of a factor depends on the level of another factor.

III. To estimate the interaction effect between nickel and gold, obtain the interaction column by multiplying the nickel column with the gold column as in Table 10.2.

$$\text{Effect of nickel*gold interaction} = \frac{64 + 52}{2} - \frac{58 + 75}{2}$$

$$= 58.0 - 66.5$$

$$= -8.5$$

Table 10.2 Interactions table for nickel and gold effects

| Design point | Coded units | | | Strength |
	Nickel	Gold	Nickel*Gold	
1	−1	−1	1	52
2	1	−1	−1	58
3	−1	1	−1	75
4	1	1	1	64

Interpretation

By simultaneously changing the amount of nickel and gold, the net effect on average is a reduction of 8.5 units on the breaking strength.

Replication

In order to determine if the above effects are real effects or statistically significant, we must have a good estimate of the experimental error. The investigator, therefore, fully replicated the above design, obtaining a total of eight runs as shown in Table 10.3.

With the new additional data, one can obtain a refined estimate of effect for each factor under study as:

$$\text{Effect of nickel} = \frac{64 + 66 + 58 + 54}{4} - \frac{75 + 71 + 52 + 49}{4}$$

$$= 60.5 - 61.75$$

$$= -2.5$$

$$\text{Effect of gold} = \frac{64 + 66 + 74 + 71}{4} - \frac{58 + 54 + 52 + 49}{4}$$

$$= 69.0 - 53.25$$

$$= 15.75$$

Table 10.3 Replication of nickel and gold design

| Design point | Coded units | | | Response | |
	Nickel	Gold	Nickel*Gold	Replicate strength	
1	−1	−1	1	52	49
2	1	−1	−1	58	54
3	−1	1	−1	75	71
4	1	1	1	64	66

$$\text{Effect of nickel*gold interaction} = \frac{64+66+52+49}{4} - \frac{75+71+58+54}{4}$$

$$= 57.75 - 64.5$$

$$= -6.75$$

The interaction plot for the example is presented in Figure 10.7.

The Model

The final model in terms of un-coded factors is:

$$\text{Strength} = 9.0 + 1.9 * \text{Nickel} + 7.2 * \text{Gold} - 0.27 * \text{Nickel} * \text{Gold}$$

The final model in terms of coded factors is:

$$\text{Strength} = 61.125 - 0.625 * \text{Nickel} + 7.875 * \text{Gold} - 3.375 * \text{Nickel} * \text{Gold}$$

Estimate of the experimental error

The above estimates of the main effects and the interaction are subjected to errors. Therefore, by running replicates of the experiment we will be able to estimate the experimental error, and this will provide us the opportunity to interpret the effect estimates in light of the error.

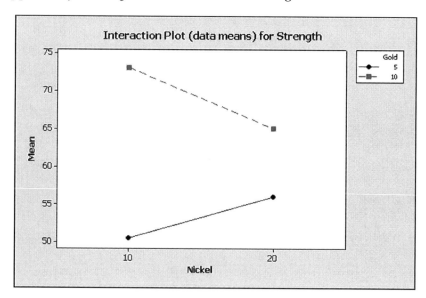

Figure 10.7 Interaction plot for example.

If we assume that the errors made in taking the observations are independent of one another, then we can estimate the experimental error by calculating the variances of the replicate observations within each design point. The Table 10.4 shows the results obtained.

If we assume that the variance is homogeneous throughout the experimental region, then we can pool the above variances. The pooled variance can be calculated as:

$$s^2_{pooled} = \frac{v_1 s_1^2 + v_2 s_2^2 + ... + v_k s_k^k}{v_1 + v_2 + ... + v_k}$$

Therefore,

$$s^2_{pooled} = \frac{(1)4.5 + (1)8 + (1)8 + (1)2}{1+1+1+1}$$

$$= \frac{22.5}{4}$$

$$= 5.625$$

Therefore, the pooled variance is equal to 5.625 with $v=4$ degrees of freedom. This pooled variance will be used to construct the confidence intervals about the estimates of effects.

Confidence intervals for the effects

The 95% confidence intervals associated with an effect can be represented as:

$$\text{Effect} \pm t_{v,0.025}\sqrt{2s^2 / n}$$

where n = total number of observations in each average: $n=4$, $v=4$
    From the t-table,

$$t_{v,0.025} = t_{5,0.025} = 2.776$$
$$s^2 = s^2 \text{ pooled} = 5.625$$

Table 10.4 Variances of replicate observations

Design point	Strength		Variance (s²)	V = d.f
1	52	49	4.5	1
2	58	54	8	1
3	75	71	8	1
4	64	66	2	1

The 95% confidence intervals for the error are:

$$\pm 2.776\sqrt{2*(5.625)/4} = \pm 4.655$$

I. The Confidence Intervals for the Effect of Nickel:
   The 95% confidence intervals for the true main effect of nickel is

$$\text{Effect} \pm 4.655 = -1.25 \pm 4.655 \ \text{ or} \left(-5.905 \text{ to } 3.405\right)$$

II. The Confidence Intervals for the Effect of Gold:
   The 95% confidence intervals for the true main effect of gold is

$$\text{Effect} \pm 4.655 = 15.75 \pm 4.655 \ \text{ or} \left(11.095 \text{ to } 20.405\right)$$

III. The Confidence Intervals for the Interaction Effect:
   The 95% confidence intervals for the interaction effect of nickel
   and gold is

$$\text{Effect} \pm 4.655 = -6.75 \pm 4.655 \ \text{ or} \left(-11.405 \text{ to } -2.095\right)$$

Conclusion

From the above analysis, we can conclude that the main effect of gold is statistically significant at $\alpha=0.05$. In addition, the interaction between nickel and gold is statistically significant at $\alpha=0.05$. This is because their confidence intervals do not contain zero. However, the main effect of nickel is not statistically significant at $\alpha=0.05$ because its confidence intervals contain zero.

Interpretation

Increasing the amount of gold from 5% to 10% increases the breaking strength by 15.75 units. The practical significance of this 15.75 units needs to be considered further. If this is of practical significance, another experiment on percent gold in a different range with some center points may be considered. This new experiment should explore the new range of the percent gold at different levels of the percent nickel since a significant interaction exists between nickel and gold.

Experimenting with three factors: 2^3 factorial design

Conjecture

We wish to study the effects of webspeed, Voltage, and WebGap on the surface roughness of a plating material. It is required that the experiment be capable of estimating all main effects as well as all interactions. Our main objective is to minimize roughness.

Design

A two-level 2^3 full factorial design with some center point replicates will allow us to investigate the three main effects, three two-factor interactions, and one three-factor interaction. The center points will serve two purposes:

- It enables us to test for curvature effect
- To obtain experimental errors if replicated

Due to the limited resources, a full 2^3 design with four center points will be considered. These four center points will give us three degrees of freedom for our estimate of experimental error. For all practical purposes, three degrees of freedom should be considered the minimum for error degrees of freedom. The conditions for the factors are shown in Table 10.5. The design matrix in standard order is presented in Figure 10.8.

Table 10.5 Factor conditions for plating study

	Low	Center	High
	(−)	(0)	(+)
Webspeed	20	30	40
Voltage	20	25	30
Web Gap	10	20	30

	Speed (S)	Voltage (V)	Gap (G)	S*V	S*G	V*G	S*V*G	Roughness
1.	−1	−1	−1	1	1	1	−1	30
2.	1	−1	−1	−1	−1	1	1	19
3.	−1	1	−1	−1	1	−1	1	37
4.	1	1	−1	1	−1	−1	−1	19
5.	−1	−1	1	1	−1	−1	1	21
6.	1	−1	1	−1	1	−1	−1	18
7.	−1	1	1	−1	−1	1	−1	34
8.	1	1	1	1	1	1	1	20
9.	0	0	0	0	0	0	0	24
10.	0	0	0	0	0	0	0	22
11.	0	0	0	0	0	0	0	23
12.	0	0	0	0	0	0	0	21

Figure 10.8 Design matrix for plating study.

The design in Figure 10.8 is randomized and run. This design is represented geometrically as a cube shown in Figure 10.9.

Analysis

The results obtained for the example is provided in Table 10.6. The ANOVA table is shown in Tables 10.7 through 10.9.

Final equation in terms of coded variables is shown below:

$$Roughness = 24.750 - 5.750 * A + 2.750 * B - 1.500 * C - 2.250 * A * B$$

$$+1.500 * A * C + 1.000 * B * C - 0.500 * A * B * C$$

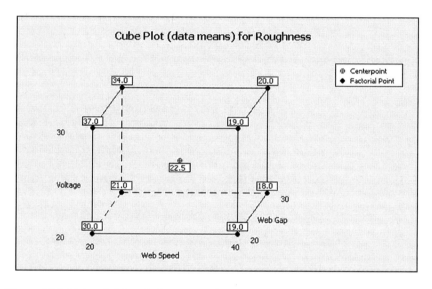

Figure 10.9 Cube plot for plating example.

Table 10.6 Experimental results for plating example part (a)

Variable	Coefficient	Standardized Effect	Sum of Squares
Overall Average	24.00	—	—
A	−5.75	−11.50	264.50
B	2.75	5.50	60.50
C	−1.50	−3.00	18.00
AB	−2.25	−4.50	40.50
AC	1.50	3.00	18.00
BC	1.00	2.00	8.00
ABC	−0.50	−1.00	2.00
Center Point	−2.25	—	13.50

Table 10.7 ANOVA result for plating example part (b)

Source	Sum of Squares	df	Square	Mean Value	F Prob > F
Model	411.50	7	58.79	35.27	0.0070
Curvature	13.50	1	13.50	8.100	0.0653
Residual	5.00	3	1.67		
Pure Error	5.00	3	1.67		
Corr Total	430.00	11			

Root MSE = 1.291
R-Squared = 0.9880
Adjusted R-Squared = 0.9600
C.V. = 5.38

Table 10.8 Experimental results for plating example part (c)

Variable	Coefficient Estimate	df	Standard Error	t for H0 Coefficient = 0	Prob > |t|
Intercept	24.750	1	0.456		
A	−5.750	1	0.456	−12.60	0.0011
B	2.750	1	0.456	6.025	0.0092
C	−1.500	1	0.456	−3.286	0.0462
AB	−2.250	1	0.456	−4.930	0.0160
AC	1.500	1	0.456	3.286	0.0462
BC	1.000	1	0.456	2.191	0.1162
ABC	−0.500	1	0.456	−1.095	0.3534
Center Point	−2.250	1	0.791	−2.846	0.0653

Final equation in terms of un-coded variables:

$$\text{Roughness} = 31.500 - 0.250 * \text{WebSpeed} + 0.900 * \text{Voltage}$$
$$- 1.850 * \text{WebGap} - 0.025 * \text{WebSpeed} * \text{Voltage}$$
$$+ 0.040 * \text{WebSpeed} * \text{WebGap} + 0.050 * \text{Voltage} * \text{WebGap}$$
$$- 0.001 * \text{WebSpeed} * \text{Voltage} * \text{WebGap}$$

Complete analysis of the example indicates that high webspeed, low voltage, and high speed can be used to minimize surface roughness.

Table 10.9 Analysis of variance – ANOVA

Source	DF	Adj SS	Adj MS	F-Value	P-Value
Analysis of Variance					
Model	16	0.016408	0.001026	4.62	0.116
Linear	4	0.005446	0.001361	6.13	0.084
Air Flow	1	0.001190	0.001190	5.36	0.104
Zone 1 Temp	1	0.000196	0.000196	0.88	0.417
Zone 2 Temp	1	0.001089	0.001089	4.91	0.114
Web Speed	1	0.002970	0.002970	13.38	0.035
2-Way Interactions	6	0.005279	0.000880	3.96	0.143
Air Flow*Zone 1 Temp	1	0.001225	0.001225	5.52	0.100
Air Flow*Zone 2 Temp	1	0.000324	0.000324	1.46	0.314
Air Flow*Web Speed	1	0.000462	0.000462	2.08	0.245
Zone 1 Temp*Zone 2 Temp	1	0.000420	0.000420	1.89	0.263
Zone 1 Temp*Web Speed	1	0.000144	0.000144	0.65	0.480
Zone 2 Temp*Web Speed	1	0.002704	0.002704	12.18	0.040
3-Way Interactions	4	0.002236	0.000559	2.52	0.237
Air Flow*Zone 1 Temp*Zone 2 Temp	1	0.000240	0.000240	1.08	0.375
Air Flow*Zone 1 Temp*Web Speed	1	0.000121	0.000121	0.55	0.514
Air Flow*Zone 2 Temp*Web Speed	1	0.001369	0.001369	6.17	0.089
Zone 1 Temp*Zone 2 Temp*Web Speed	1	0.000506	0.000506	2.28	0.228
4-Way Interactions	1	0.000870	0.000870	3.92	0.142
Air Flow*Zone 1 Temp*Zone 2 Temp*Web Speed	1	0.000870	0.000870	3.92	0.142
Curvature	1	0.002576	0.002576	11.61	0.042
Error	3	0.000666	0.000222		
Total	19	0.017074			

Table 10.10 Model summary

S	R-sq	R-sq(adj)
0.0148997	96.10%	75.30%

Coded Coefficients					
Term	Effect	Coef	SE Coef	T-Value	P-Value
Constant		1.97538	0.00372	530.31	0.000
Air Flow	−0.01725	−0.00862	0.00372	−2.32	0.104
Zone 1 Temp	0.00700	0.00350	0.00372	0.94	0.417
Zone 2 Temp	0.01650	0.00825	0.00372	2.21	0.114
Web Speed	0.02725	0.01362	0.00372	3.66	0.035
Air Flow*Zone 1 Temp	−0.01750	−0.00875	0.00372	−2.35	0.100
Air Flow*Zone 2 Temp	−0.00900	−0.00450	0.00372	−1.21	0.314
Air Flow*Web Speed	−0.01075	−0.00537	0.00372	−1.44	0.245
Zone 1 Temp*Zone 2 Temp	0.01025	0.00512	0.00372	1.38	0.263
Zone 1 Temp*Web Speed	0.00600	0.00300	0.00372	0.81	0.480
Zone 2 Temp*Web Speed	0.02600	0.01300	0.00372	3.49	0.040
Air Flow*Zone 1 Temp*Zone 2 Temp	−0.00775	−0.00387	0.00372	−1.04	0.375
Air Flow*Zone 1 Temp*Web Speed	−0.00550	−0.00275	0.00372	−0.74	0.514
Air Flow*Zone 2 Temp*Web Speed	−0.01850	−0.00925	0.00372	−2.48	0.089
Zone 1 Temp*Zone 2 Temp*Web Speed	0.01125	0.00563	0.00372	1.51	0.228
Air Flow*Zone 1 Temp*Zone 2 Temp*Web Speed	−0.01475	−0.00738	0.00372	−1.98	0.142
Ct Pt		□0.02838	0.00833	□3.41	0.042

Experimenting with four factors: 2^4 factorial design

Conjecture

We wish to study the effects of four factors on caliper measurements in mls. The factors under consideration are Air Flow, Zone 1 Temp, Zone 2 Temp, and webspeed. It is required that the experiment be capable of estimating all main effects as well as all interactions. Our main objective is to have the caliper at a target of 1.97 mls.

Design

A two-level 2^4 full factorial design with some four center point replicates will allow us to investigate the four main effects, six two-factor interactions, four three-factor interactions, and one four-factor interaction. The four center points will enable us to test for curvature effect as well as obtaining our experimental errors.

Alias Structure

Factor Name
    A Air Flow
    B Zone 1 Temp
    C Zone 2 Temp
    D webspeed

Aliases
    I
    A
    B
    C
    D
    AB
    AC
    AD
    BC
    BD
    CD
    ABC
    ABD
    ACD
    BCD
    ABCD

Due to the limited resources, a full 2^4 design with four center points was considered rather than replicating the whole full design that would have resulted in a total of 32 experimental runs. These four center points will give us three degrees of freedom needed for our estimate of experimental error. For all practical purposes, three degrees of freedom should be considered minimum for error degrees of freedom. The design matrix obtained in standard order together with the conditions for the factors are presented below in Figure 10.10.

The design in Figure 10.10 is randomized and run in random order. This design is represented geometrically as a cube in Figure 10.11.

Analysis

The results obtained are provided below. The ANOVA table is shown in Table 10.9.

The ANOVA table shows that the main effect of the webspeed with $p = 0.035$ is statistically significant in controlling the caliper. In addition, the interaction between Zone 2 Temp and webspeed is also statistically significant ($p = 0.04$). This means that we cannot look at the effect of webspeed alone in controlling the caliper. Therefore, we will have to look at the effect of the webspeed at each level of the interacting factor which is the Zone 2 Temp (Table 10.10).

Std Order	Run Order	Center Point	Blocks	Air Flow	Zone 1 Temp	Zone 2 Temp	Web Speed	Caliper (mils)
1	18	1	1	2	165	165	75	1.964
2	12	1	1	4	165	165	75	1.967
3	6	1	1	2	215	165	75	1.985
4	15	1	1	4	215	165	75	1.950
5	7	1	1	2	165	215	75	1.953
6	4	1	1	4	165	215	75	1.961
7	8	1	1	2	215	215	75	1.958
8	5	1	1	4	215	215	75	1.956
9	11	1	1	2	165	165	125	1.972
10	3	1	1	4	165	165	125	1.972
11	13	1	1	2	215	165	125	1.964
12	2	1	1	4	215	165	125	1.963
13	16	1	1	2	165	215	125	1.998
14	1	1	1	4	165	215	125	1.988
15	19	1	1	2	215	215	125	2.078
16	9	1	1	4	215	215	125	1.977
17	20	0	1	3	190	190	100	1.953
18	17	0	1	3	190	190	100	1.938
19	14	0	1	3	190	190	100	1.965
20	10	0	1	3	190	190	100	1.932

Figure 10.10 Full 2^4 design matrix.

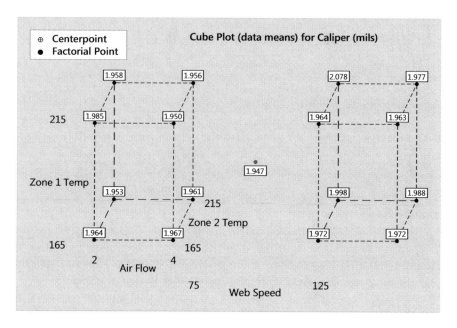

Figure 10.11 Cube plot for caliper (mls).

Regression equation in un-coded units
Caliper (mils) = -2.37 + 1.229 Air Flow + 0.0280 Zone 1 Temp
+ 0.0230 Zone 2 Temp
+ 0.0495 Web Speed 0.00770 Air Flow*Zone 1 Temp
                    - 0.00649 Air Flow*Zone 2 Temp -
  0.01361 Air Flow*Web Speed
                    - 0.000151 Zone 1 Temp*Zone 2 Temp -
  0.000319 Zone 1 Temp*Web Speed
              - 0.000272 Zone 2 Temp*Web Speed
+ 0.000041 Air Flow*Zone 1 Temp*Zone 2 Temp
              + 0.000085 Air Flow*Zone 1 Temp*Web Speed
              + 0.000075 Air Flow*Zone 2 Temp*Web Speed
              + 0.000002 Zone 1 Temp*Zone 2 Temp*Web Speed
- 0.000000 Air Flow*Zone 1 Temp*Zone 2 Temp*Web Speed -
  0.02837 Ct Pt

Effects pareto for caliper (mils)
The Pareto plot of the factor effects in Figure 10.12 also confirmed that the
main effect of the webspeed as well as the interaction between webspeed
and Zone 2 Temp are the statistically significant factors for controlling the
caliper as previously shown from the ANOVA table.

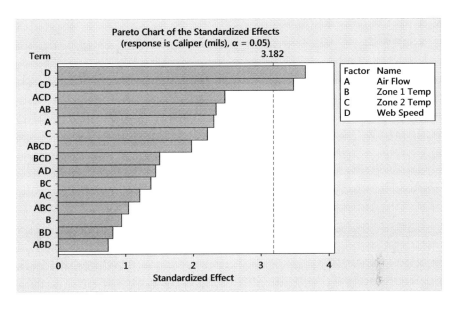

Figure 10.12 Pareto chart of effects.

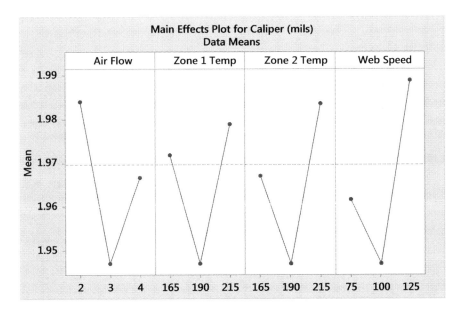

Figure 10.13 Main effects plot for caliper (mils).

Main effects plot for caliper (mils)

The main effects plot shows the significance of adding the center points which enabled us to assess the quadratic effect of each factor. The center points showed statistically significant curvature effect (p=0.042) as can be seen from the main effects plot below. This means that a design that is capable of estimating a second order model is needed, such as a response surface design which will be presented in the next chapter. The current design is only good for a linear model (Figure 10.13).

Interaction effects plot for caliper (mils)

Since the ANOVA shows that there is a significant interaction effect between webspeed and Zone 2 Temp, the effect of webspeed alone cannot be used in optimizing the caliper target of 1.97. Instead, the effect of webspeed has to be assessed at each level of Zone 2 Temp. It can be seen from the interaction plot below that at high webspeed of 125 and high Zone 2 Temp of 215, the caliper is at the maximum of 2.01. Similarly, when the webspeed is still at high level of 125 but the Zone 2 Temp is at a low level of 165, the caliper is about 197 mls. Therefore, a high webspeed of 125 and low Zone 2 Temp will enable us to achieve our caliper target of 197 mls. In addition, a low webspeed of 75 and a low Zone 2 Temp of 165 will also enable us to achieve our caliper target of 1.97 mls. As shown previously through a statistically significant center point, a response surface design is needed to be able to achieve global optimization for the caliper (Figure 10.14).

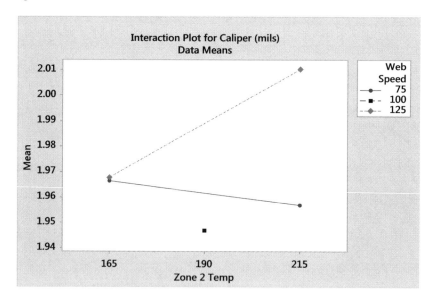

Figure 10.14 Interaction plot for caliper (mils).

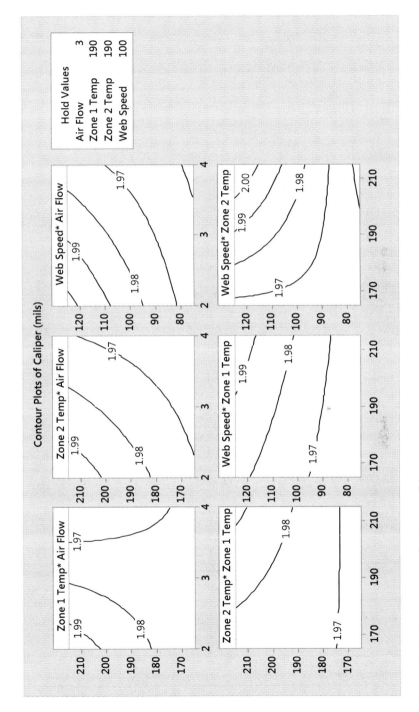

Figure 10.15 Contour plot of caliper (mils).

Contour plot of the caliper (mils)

The linear model obtained can be used to generate the contour plots. The result obtained is provided below for all combinations of the factors under study. This result shows that there are many optimal operating regions that we can use to obtain the target caliper of 1.97 (Figure 10.15).

Bibliography

Ayeni, B.J. (1999). *Application of Dual-Response Optimization for Moving Web Processes*, 3M Internal Publication, IT Statistical Consulting, 3M Company, St. Paul, MN.

Badiru, A.B. and Ayeni, B.J. (1993), *Practitioner's Guide to Quality and Process Improvement*, Chapman & Hall, London.

Box G.E.P, Hunter W.G, and Hunter J.S. (1978). *Statistics for Experimenters*, John Wiley & Sons, New York.

Fisher, R.A. (1935). *The Design of Experiments*, 8th edn, Oliver & Boud, Edinburgh.

Taguchi, G (1986). *Introduction to Quality Engineering: Designing Quality into Products and Processes*, Kraus International Publications, White Plains, NY.

Taguchi, G. and Wu, Y. (1985). *Introduction to Off-line Quality Control*, Central Japan Quality Control Association, Nagaya, Japan.

Yates, F., (1935) Complex Experiments. *Supplement to the Journal of the Royal Statistical Society*, II(2), 181–247.

chapter eleven

Fractional factorial experiments

Introduction

Two-level factorial designs represent an efficient way of investigating the effects of several independent variables on a response of interest. However, each time the number of variables or factors increases by one, the size of the experiment increases by a factor of two, and once beyond three or four variables the number of experimental runs to be made soon becomes overwhelming as presented in Table 11.1 below.

For many practical situations, it may be impossible to collect all the observations required by a full factorial experiment. In such cases, fractional factorial experiments are used. In a fractional factorial experiment, only a fraction of the treatment replicates are run. The advantages of fractional factorial experiments include the following:

- Lower cost of experimentation
- Reduced time for experimentation
- Efficiency of analysis

When the number of factors, k, is greater than 5 (k > 5), then the number of runs required for a full factorial experiment will be impractical for most industrial applications. For example, from the above 2^k factorial design table, it will require a total of 32,768 experimental runs to study the effects of 15 variables in a standard full factorial environment which is not economically feasible in an industrial environment. However, in some industries, such as semiconductor industry, where computer simulations are used in the design phase of certain products, the number of runs will be of little concern. Whereas, in most other industries, such as chemical, petrochemical, process, paper and pulp, as well as parts industries, where a large number of factors must be examined, it will often be very desirable to reduce the number of runs in an experiment by taking a fraction of the full 2^k factorial design.

In a 2^{k-p} fractional factorial design there are:

- 2 levels of each factor under consideration
- k number of factors to be studied
- 2^{k-p} that can be estimated including the overall mean
- 2^{k-p} minimum number of experimental runs
- p number of independent generators
- 2^{k-p} number of words in defining relations (including I)

Table 11.1 2^k factorial design table

K	Number of runs
2	4
3	8
4	16
5	32
6	64
7	128
8	256
9	512
10	1024
11	2048
12	4096
13	8192
14	16384
15	32768

The requirement is that $2^{k-p} > k$

p = degree of fractionation
p = 1 (half fraction)
p = 2 (quarter fraction)
2^{k-p} = number of distinct conditions in the cube portion of a design.

For example, a one-half of a 2^3 factorial design is referred to as 2^{3-1} fractional factorial design.

The disadvantages of fractional designs involve the loss of one or more of the interaction effects that can be studied in a full factorial design. Also, the design of a fractional factorial experiment can be complicated since it may be difficult to select the treatment combinations to be used.

Fractional factorial designs are denoted as follows:

1/2 fractional design: one-half of complete factorial experiment
1/4 fractional design: one-fourth of complete factorial experiment
1/8 fractional design: one-eighth of complete factorial experiment

A 2^4 factorial design

Conjecture

We want to investigate all combinations of two levels of each of four factors, A, B, C, and D, and obtain estimates of all effects including all interactions. We may wish to include some center points or replicate points to obtain an estimate of the experimental error.

Design

The design can be setup as shown in Figure 11.1.

The analysis of the above design will provide uncorrelated and independent estimates of the following:

- The overall average of the response.
- The main effects due to each factor, A, B, C, D.
- The estimates of six two-factor interaction effects, AB, AC, AD, BC, BD, CD.
- The estimates of four three-factor interaction effects, ABC, ABD, ACD, BCD.
- The estimate of one four-factor interaction effect, ABCD.

Full factorial designs can be generated for any number of factors, k. However, it should be noted that the number of runs needed for a full factorial design increases rapidly with increasing values of k.

Conjecture

An investigator wishes to investigate the effects of four factors (k=4) using only eight runs.

Design Point	A	B	C	D	AB	AC	AD	BC	BD	CD	ABC	ABD	ACD	BCD	ABCD
1	−	−	−	−	+	+	+	+	+	+	−	−	−	−	+
2	+	−	−	−	−	−	−	+	+	+	+	+	+	−	−
3	−	+	−	−	−	+	+	−	−	+	+	+	−	+	−
4	+	+	−	−	+	−	−	−	−	+	−	−	+	+	+
5	−	−	+	−	+	−	+	−	+	−	−	−	+	+	−
6	+	−	+	−	−	+	−	−	+	−	−	+	−	+	+
7	−	+	+	−	−	−	+	+	−	−	−	+	+	−	+
8	+	+	+	−	+	+	−	+	−	−	+	−	−	−	−
9	−	−	−	+	+	+	−	+	−	−	−	+	+	+	−
10	+	−	−	+	−	−	+	+	−	−	+	−	−	+	+
11	−	+	−	+	−	+	−	−	+	−	+	−	+	−	+
12	+	+	−	+	+	−	+	−	+	−	−	+	−	−	−
13	−	−	+	+	+	−	−	−	−	+	+	+	−	−	+
14	+	−	+	+	−	+	+	−	−	+	−	−	+	−	−
15	−	+	+	+	−	−	−	+	+	+	−	−	−	+	−
16	+	+	+	+	+	+	+	+	+	+	+	+	+	+	+

Figure 11.1 Design setup for 2^4 factorial design.

Design

A 2^{4-1} fractional factorial design of 8 runs can be set up as illustrated below:

Procedure

I. Set up a full 2^3 design of 8 runs as shown in Figure 11.2.
II. Assign the fourth factor D to the highest order interaction.

$$D = ABC$$

D = ABC is known as generator

III. Generate the + and − column for D by multiplying columns A, B, and C together to obtain the layout in Figure 11.3. Note that the levels of factor D are the products of the levels of factors A, B, and C.

Design Point	A	B	C
1	−	−	−
2	+	−	−
3	−	+	−
4	+	+	−
5	−	−	+
6	+	−	+
7	−	+	+
8	+	+	+

Figure 11.2 Setup of a full 2^3 design.

Design Point	A	B	C	D=ABC
1	−	−	−	−
2	+	−	−	+
3	−	+	−	+
4	+	+	−	−
5	−	−	+	+
6	+	−	+	−
7	−	+	+	−
8	+	+	+	+

Figure 11.3 Layout for design points.

IV. Obtain the defining relation (I) by multiplying both sides of the generator by D as:

$$D * D = ABC * D$$

$$I = ABCD \left(A \text{ resolution IV design} \right)$$

The defining relation will be used to determine the resolution of a design and to generate the confounding patterns of the effects.

Design resolution

The resolution of a design is the number of letters in the smallest word of the defining relation. For example, in the design shown in Figure 11.1, the defining relation I = ABCD has one word which is ABCD, and the number of letters is equal to 4, hence a resolution IV design. In a resolution IV design, the main effects are confounded with three-factor and higher order interactions, and two-factor interactions are confounded with each other and higher order interactions. Similarly, a resolution III design has the main effects confounded with two-factor and higher order interactions. A 2^{5-2} design of 8 runs is of resolution III. An extremely useful design is a 2^{5-1} design of resolution V.

This design is used to investigate five factors in only 16 runs and has the power to estimate the main effects clear of any other factors and the two-factor interactions clear of any other factors as well if three-factor and higher order interactions are assumed to be negligible. Additional information on design resolution can be found in Box, Hunter, and Hunter (1978), Ayeni (1991), as well as in many other experimental design books and papers.

The defining relation (I) is the column of $+1$. Any factor multiplied by itself gives I. For example:

$$A * A = I, B * B = I, AB * AB = AABB = II = I,$$

$$ABC * ABC = AABBCC = III = I$$

Other operators which are not equal to I are:

$$A * AB = B = IB = B, AB * BCE = AICE = ACE$$

V. Use the defining relation I = ABCD to generate the effects or confounding patterns as follows:
   1. To obtain the confounding patterns for main effect A, multiply both sides of the defining relation by A as:

$$A * I = A * ABCD = BCD.$$

Therefore, A = BCD. This means that when we estimate the effect of factor A, we are not only estimating the effect of A but also the effect of the three-factor interaction BCD. This is known as confounding.

Confounding

Confounding occurs when the effects of two or more factors cannot be separated. In the above example, where A = BCD, we are really estimating the sum of two effects A + BCD.

2. Similarly, to obtain any two-factor interaction, say BC, multiply both sides of defining relation by BC as:

$$BC * I = BC * ABCD = AD.$$

Therefore, BC = AD, that is the effects of BC and AD are confounded with each other. Thus, estimating the effect of BC implies that we are really estimating the sum of BC + AD.

The complete confounding patterns are shown in Figure 11.4.

From the above main effects, we can see that when we believe we are estimating the main effects, we are actually estimating the sum of the main effects and the three-factor interactions. However, since three-factor interactions and higher order interactions are generally assumed to be negligible or nonexistence, we can obtain estimates of all the main effects clear of all other effects. Although all the two-factor interactions are confounded with each other, this is the price we pay in running eight experiments rather than 16 experiments. Unless certain two-factor interactions are known not to exist, it will be necessary to run another half fraction if each two-factor effect is to be estimated clear of all other effects. Interested readers should refer to Hicks (1982) for further details on factorial designs and fractional factorial designs.

Saturated designs

Saturated designs are designs that can be used to investigate n-1 factors in *n* number of runs. For example, one can study the effects of 15 factors using only 16 runs. In fact, it is also possible to investigate the effects of 31 factors using 32 runs. These designs are extremely useful in screening applications as well as in situations where main effects are believed to dominate over two-factor and higher order interactions. All saturated designs are of resolution III. This means that the main effects are confounded with two-factor and higher order interactions. It is, therefore, extremely important to initially screen for factors that are critical to the response of interest, since only a few of these factors exist, and later conduct a more thorough

Effect	2^{4-1}	Confounding Patterns
A	IV	A + BCD
B		B + ACD
C		C + ABD
D		D + ABC
AB		AB + CD
AC		AC + BD
BC		BC + AD

Figure 11.4 Confounding patterns for factorial design example.

Number of Factors	Number of Runs	Type of Designs
3	4	2^{3-1}
7	8	2^{7-4}
15	16	2^{15-11}
31	32	2^{31-26}
63	64	2^{63-57}

Figure 11.5 Examples of saturated designs.

investigation of those factors identified through a full factorial design or a central composite design. Examples of saturated designs are shown in Figure 11.5.

Example of a saturated design

A 2^{7-4} fractional factorial design can be used to investigate $k=7$ factors with only 8 runs. For this design, the number of fractions is $p=4$. In any saturated design, all possible interactions are used up in building the generators. Since $p=4$ in a 2^{7-4} design, there will be a total of 4 generators. The design set up is provided in Figure 11.6.

Procedure Step I

Set up a full 2^3 design of 8 runs (see Figure 11.6).

Procedure Step II

Assign the remaining four variables (D, E, F, G) to all the interactions to obtain the 4 generators.

Design Point	A	B	C
1	–	–	–
2	+	–	–
3	–	+	–
4	+	+	–
5	–	–	+
6	+	–	+
7	–	+	+
8	+	+	+

Figure 11.6 Design setup for saturated design.

$$D = AB\ E = AC\ F = BC \text{ and } G = ABC \text{ (These are the generators)}$$

Procedure Step III

Generate the + and – columns for the generators as shown in Figure 11.7.

Procedure Step IV

Generate the defining relation (I) as presented below. Notice that the smallest word has three letters, hence a resolution III design.

$$I = ABD = ACE = BCF = ABCG = BCDE = ACDF = CDG = ABEF = BEG$$

$$= AFG = DEF = ADEG = BDFG = CEFG = ABCDEF$$

Procedure Step V

The confounding patterns can be generated as before by multiplying the factors on both sides of the defining relations. The confounding patterns

Design Point	A	B	C	D=AB	E=AC	F=BC	G=ABC
1	–	–	–	+	+	+	–
2	+	–	–	–	–	+	+
3	–	+	–	–	+	–	+
4	+	+	–	+	–	–	–
5	–	–	+	+	–	–	+
6	+	–	+	–	+	–	–
7	–	+	+	–	–	+	–
8	+	+	+	+	+	+	+

Figure 11.7 Generation of +/– columns for saturated design.

Effects	2^{7-4}	Confounding Patterns
A	III	A+BD+CE+FG
B		B+AD+CF+EG
C		C+AE+BF+DG
AB		AB+D+CG+EF
AC		AC+E+BG+DF
BC		BC+F+AG+DE
ABC		CD+BE+AF+G

Figure 11.8 Confounding patterns for saturated design.

are provided in Figure 11.8 after three-factor and higher order interactions have been deleted.

Response surface methodology

Response surface methodology (RSM) involves an analysis of the prediction equation or response surface fitted to a set of experimental data. Response surface strategies can be classified into two categories. These are single phase and double phase strategies.

Single phase strategy

This strategy requires running a full factorial design plus center points and star points to fit a second order response surface. This is shown graphically in Figure 11.9.

Double phase strategy

This strategy requires the initial running of a full factorial design with some center points. Analyze the data by fitting a first order model, then

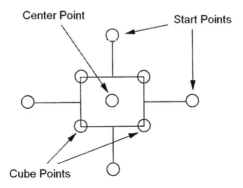

Figure 11.9 Single-phase response surface strategy.

test for lack of fit, and use the center points to test for the effects of curvature. If there is a significant lack of fit or if the quadratic effect is significant or both, then proceed further by running a second design which includes the star points and additional center points. Then analyze the data from the two designs together. A double-phase response surface strategy is depicted in Figure 11.10.

The selection of which design phase to consider depends on several factors which are discussed below:

- The Major Goal of the Experiment

If the major goal of the experiment is to optimize the process and one is considering only two to three factors to study with no other problems, then one can proceed straight with single phase strategy.

- The Cost of Running the Experiment

If the cost of running the experiment is a major concern and one is required to minimize cost as much as possible, then select a double phase strategy. By fitting a first order model first and further determine through a curvature test that a second order model is not necessary will not only save you a substantial amount of money, but will at the same time save you a fairly large amount of experimental time.

- The Number of Variables to Be Studied

If the number of variables to be studied is greater than five, then select a double phase strategy and run a fractional factorial of resolution IV or better first.

- The Time Required to Complete the Project

If longer time is required to complete each run in the experiment, then you may select a double phase strategy. You will save a lot of time if it is determined that a second-degree model is not necessary.

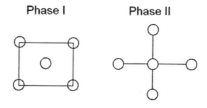

Figure 11.10 Double-phase response surface strategy.

- Type of Control Variables Under Study

If all the control variables are qualitative variables, then select double phase strategy. You only need to run a full or fractional factorial design with some replicates. No star points or center points are possible for this case.

- Prior Knowledge of the Experimenter

If the experimenter (through prior experiment or any other means) knew in advance that within the range of study, the first order model would be adequate, then select a double phase.

- Maximum Number of Runs Possible

If there is a limitation on the number of runs possible, then consider double phase strategy.

Response surface example

A process engineer has just completed a 2^{6-2} screening design where he studied 6 factors on the mineral penetration of a fiber. The response of interest is the fiber thickness. The screening experiment identified three key factors, webspeed, % Solids, and Fiber Weight. The engineer decided to determine the optimum operating conditions of these critical factors that can be used to achieve a minimum thickness of 0.40.

Objective

To determine control handles that will achieve a target thickness of 0.40 or better.

Design

A three-factor central composite design as shown below is selected. The ranges of interest to be studied are shown in Table 11.2. The resulting design and the response are presented in Table 11.3. Table 11.4 shows the analysis of variance (ANOVA) results for thickness for a three-factor study.

Table 11.2 Three-factor central composite design

Factors	−1.633	−1	0	1	1.633
Webspeed	43.67	50	60	70	76.33
% Solids	36.83	40	45	50	53.16
Fiber Weight	16.83	20	25	30	33.16

Table 11.3 Data for response surface example

Obs. no	Run order	Block	Webspeed X1	% Solids X2	Fiber weight X3	Thickness (Response)	Design ID
1	6	1	50.000	40.000	20.000	0.320	1
2	1	1	70.000	40.000	30.000	0.336	2
3	5	1	50.000	50.000	30.000	0.361	3
4	3	1	70.000	50.000	20.000	0.399	4
5	2	1	60.000	45.000	25.000	0.404	5
6	4	1	60.000	45.000	25.000	0.380	6
7	8	2	50.000	40.000	30.000	0.321	7
8	12	2	70.000	40.000	20.000	0.356	8
9	9	2	50.000	50.000	20.000	0.350	9
10	11	2	70.000	50.000	30.000	0.404	10
11	10	2	60.000	45.000	25.000	0.353	11
12	7	2	60.000	45.000	25.000	0.375	12
13	14	3	43.670	45.000	25.000	0.353	13
14	16	3	76.330	45.000	25.000	0.373	14
15	19	3	60.000	36.835	25.000	0.342	15
16	13	3	60.000	53.165	25.000	0.441	16
17	17	3	60.000	45.000	16.835	0.361	17
18	15	3	60.000	45.000	33.165	0.348	18
19	18	3	60.000	45.000	25.000	0.378	19
20	20	3	60.000	45.000	25.000	0.374	20

Analysis

Analysis of the data in Table 11.3 yields the following results:
    Estimated effects for thickness for a three-factor study

    average = 0.3776
    A:X1 = 0.0264
    B:X2 = 0.0514
    C:X3 = −0.0036
    AB = 0.0102
    AC = −0.0068
    BC = 0.0088
    AA = −0.0166
    BB = 0.0048
    CC = −0.0230
    Block 1 = 0.001400
    Block 2 = −0.012166
    Block 3 = 0.010666

    Standard error estimated from total error with 8 d.f. (t = 2.30665)

Table 11.4 ANOVA for thickness for three-factor study

Independent Variable	Coefficient Estimate	df	Standard Error	*t* for H0 Coeff. = 0	Prob > |*t*|
Intercept	0.3776	1	0.0066	57.6000	
Block 1	0.0007				
Block 2	−0.0061				
Block 3	0.0053				
A: Webspeed	0.0132	1	0.0044	2.9930	0.0173
B: % Solids	0.0257	1	0.0044	5.8380	0.0004
C: Fiber Weight	−0.0018	1	0.0044	−0.4128	0.6906
AA	−0.0083	1	0.0044	−1.8850	0.0962
BB	0.0024	1	0.0044	0.5315	0.6095
CC	−0.0115	1	0.0044	−2.6050	0.0314
AB	0.0051	1	0.0057	0.9017	0.3935
AC	−0.0034	1	0.0057	−0.5938	0.5690
BC	0.0044	1	0.0057	0.7698	0.4635
Total error	0.00207	8			
Total (corr.)	0.01682	19			

R-squared = 0.8737

R-squared (adj. for df) = 0.7316

Regression coefficients for thickness for a three-factor response surface study

Constant = 0.3776
Block 1 = 0.0007
Block 2 = −0.0061
Block 3 = 0.0053
A: Webspeed = 0.0132
B: %Solids = 0.0257
C: Fiber Weight = −0.0018
AA = −0.0083
BB = 0.0024
CC = −0.0115
AB = 0.0051
AC = −0.0034
BC = 0.0044

Figure 11.11 shows the Pareto chart for the response (thickness). The chart indicates the relative contributions of the effects from the three-factor interactions. Figure 11.12 shows the response surface with respect to X2 and X3. Figure 11.13 shows the response surface with respect to X1 and X3. Figure 11.14 shows the contour surface with respect to X1 and X2. Figure 11.15 shows the contour surface with respect to X1 and X3. Figure 11.16 shows the response surface with respect to X1 and X2.

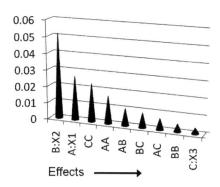

Figure 11.11 Pareto chart for the response surface analysis.

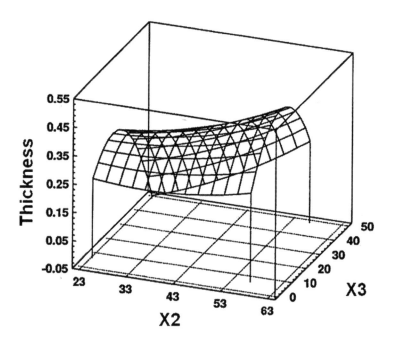

Figure 11.12 Response surface with respect to X2 and X3.

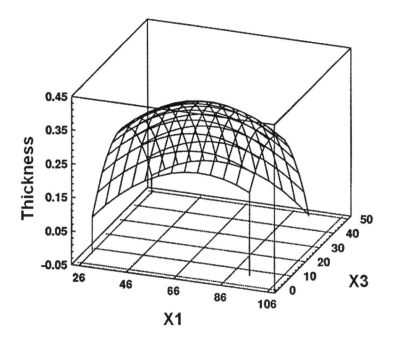

Figure 11.13 Response surface with respect to X1 and X3.

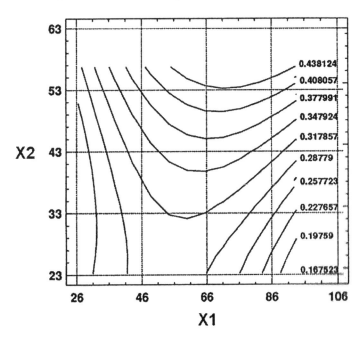

Figure 11.14 Contour surface with respect to X1 and X2.

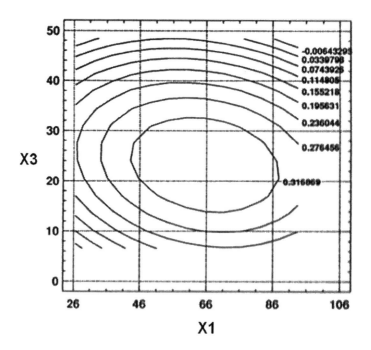

Figure 11.15 Contour surface with respect to X1 and X3.

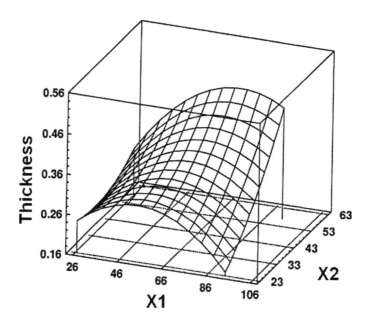

Figure 11.16 Contour surface with respect to X1 and X2.

The final equation in terms of actual factors is:

$$\text{Thickness} = 0.0509 + 0.008396(\text{Webspeed}) - 0.01385(\%\text{Solids})$$

$$+ 0.01886(\text{Fiberweight}) - 0.0000833(\text{Webspeed}^2)$$

$$+ 0.00009404(\%\text{Solid}^2) - 0.000461(\text{Fiberweight}^2)$$

$$+ 0.0001025(\text{Webspeed} * \%\text{Solids})$$

$$- 0.0000675(\text{Webspeed} * \text{Fiberweight})$$

$$+ 0.000175(\%\text{Solids} * \text{Fiberweight}).$$

Central composite designs

1. Two-Factor Central Composite Design – Design orthogonally blocked and rotatable. An example of this is shown in Table 11.5.
2. Two-Factor Central Composite Design – Factorial portion replicated and design orthogonally blocked and nearly rotatable as shown in Table 11.6.
3. Three-Factor Central Composite Design – Orthogonally Blocked and Nearly Rotatable as seen in the layout in Table 11.7.
4. Four-Factor Central Composite Design – Orthogonally blocked and rotatable as shown in Table 11.8. Figure 11.17 illustrates the graphical composition of a three-factor central composite design.

Table 11.5 Two-factor central composite design

Run	A	B	
1	−1	−1	**Block 1**
2	1	−1	
3	−1	1	
4	1	1	
5	0	0	
6	0	0	
7	−1.414	0	**Block 2**
8	1.414	0	
9	0	−1.414	
10	0	1.414	
11	0	0	
12	0	0	

Table 11.6 Two-factor central composite design with three blocks

Run	A	B	
1	−1	−1	**Block 1**
2	1	−1	
3	−1	1	
4	1	1	
5	0	0	
6	0	0	
7	−1	−1	**Block 2**
8	1	−1	
9	−1	1	
10	1	1	
11	0	0	
12	0	0	
13	−1.633	0	**Block 3**
14	1.633	0	
15	0	−1.633	
16	0	1.633	
17	0	0	
18	0	0	
19	0	0	
20	0	0	

Response surface optimization

Controlling processes to target as well as minimize variation have been important issues in recent years in most industrial organizations. Well-designed experiments can significantly impact product and process quality. This section of the book focuses on some practical issues commonly associated with moving web processes as well as those useful in reducing variability during industrial experimentation. The fundamental principle of model building using two-level factorial and fractional factorial designs will be covered. The practice of adding center points as well as "axial" points for detection of model curvature is explored. Some examples of industrial experiments are presented, including applications involving response surface designs for moving web type processes with machine direction (MD) and cross direction (CD), such as paper and plastic film productions. The problems and opportunities provided by simultaneously studying dispersion effects and location effects with the objective of achieving mean on target with minimum variance are investigated with real life examples. A numerical illustration with the use of contours is produced as a mechanism

Table 11.7 Layout for three-factor central composite design

Run	A	B	C	
1	−1	−1	1	**Block 1**
2	1	−1	−1	
3	−1	1	−1	
4	1	1	1	
5	0	0	0	
6	0	0	0	
7	−1	−1	−1	**Block 2**
8	1	−1	1	
9	−1	1	1	
10	1	1	−1	
11	0	0	0	
12	0	0	0	
13	−1.633	0	0	**Block 3**
14	1.633	0	0	
15	0	−1.633	0	
16	0	1.633	0	
17	0	0	−1.633	
18	0	0	1.633	
19	0	0	0	
20	0	0	0	

for process improvement. Multiple response optimization methods are presented as an approach to finding the common region for process optimization.

Applications for moving web processes

Factorial designs have been extensively used in industry over the last several years primarily because of their ability to efficiently provide valuable information about main effects as well as their interactions. In addition, factorial designs often provide the basic foundation for RSM and mixture (formulation) experiments. Response surface methodology is a statistical technique useful for building and exploring the relationship between a response variable, y, and a set of independent factors, x, which can be represented as:

$$y = f(x, \beta) + \varepsilon$$

Where x is a vector of factor settings, β is a vector of parameters including main effects and interactions and ε is a vector of random errors which are assumed to be independent with zero mean and common variance s2.

Table 11.8 Layout for four-factor central composite design

Run	A	B	C	D	
1	1	1	−1	1	**Block 1**
2	1	−1	−1	−1	
3	−1	1	−1	−1	
4	1	1	1	−1	
5	−1	−1	1	−1	
6	1	−1	1	1	
7	−1	1	1	1	
8	−1	−1	−1	1	
9	0	0	0	0	
10	0	0	0	0	
11	0	0	0	0	
12	1	1	−1	−1	**Block 2**
13	1	−1	1	−1	
14	−1	1	−1	1	
15	−1	−1	−1	−1	
16	−1	−1	1	1	
17	1	−1	−1	1	
18	−1	1	1	−1	
19	1	1	1	1	
20	0	0	0	0	
21	0	0	0	0	
22	0	0	0	0	
23	−2	0	0	0	**Block 3**
24	2	0	0	0	
25	0	−2	0	0	
26	0	2	0	0	
27	0	0	−2	0	
28	0	0	2	0	
29	0	0	0	−2	
30	0	0	0	2	
31	0	0	0	0	
32	0	0	0	0	
33	0	0	0	0	

One is usually interested in optimizing $f(x, \beta)$ over some appropriate design region. This method only approximates the true response surface with a convenient mathematical function. Therefore, for most practical situations, if an appropriate design region has been chosen, a quadratic function will usually approximate the true response surface quite well.

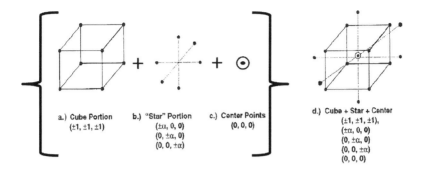

Figure 11.17 Graphical representation of central composite design.

In industry today, engineers have experienced increasing needs to develop experimental strategies that will achieve target for a quality characteristic of interest while simultaneously minimizing the variance. The classical experimental designs described above have been lacking in this area because they tend to focus solely on the mean of the quality characteristic of interest. In today's competitive manufacturing environment, engineers must be more ambitious in finding conditions where variability around the target is as small as possible in addition to meeting the target condition.

Taguchi and Wu (1985) and Taguchi (1986) have presented the need for considering the mean and variance of quality characteristics of interest. Unfortunately, Taguchi's statistical approach to this problem has drawn much criticism in the literature, Box (1985). For this reason, Vining and Myers (1990) developed a dual response approach for which one can achieve the primary goal of the Taguchi philosophy. This enables us to obtain a target condition on the mean while minimizing the variance, within a RSM framework. The example used by Vining and Myers (1990) was taken from the Box, Hunter, and Hunter (1978) book. Therefore, this chapter attempts to apply this dual response approach to real life practical examples and problems.

Dual response approach

Vining and Myers (1990) used the dual response problem formulation developed by Myers and Carter (1973). In their development, the investigator is assumed to be seeking to optimize two responses. We shall let y_p represent the response of primary interest, and y_s represents the response of secondary interest. They further assumed that these responses might be modeled by the equations below:

$$y_p = \beta_o + \sum_{i=1}^{k} \beta_i x_i + \sum_{i=1}^{k} \beta_{ii} x_i^2 + \sum \sum_{i<j} \beta_{ij} x_i x_j + \varepsilon_p$$

$$y_s = \gamma_o + \sum_{i=1}^{k} \gamma_i x_i + \sum_{i=1}^{k} \gamma_{ii} x_i^2 + \sum \sum_{i<j}^{k} \gamma_{ij} x_i x_j + \varepsilon_s,$$

where the β's and the γ's represent the unknown coefficients, and ε_p and ε_s are the random errors.

The practical significance of this approach is to optimize the primary response subject to an appropriate constraint on the value of the secondary response. The decision as to which response to make as the primary response depends solely on the ultimate goal of the experiment. By considering Taguchi's three situations:

1. "Target value is best," which means keeping mean at a specified target value while minimizing variance. This requires that the variance be the primary response.
2. "The larger the better" means making the mean as large as possible while controlling variance
3. "The smaller the better" means making the mean as small as possible while controlling the variance. These last two require the mean to be the primary response.

Case study of application to moving webs

In this application, a process capability experiment was performed on a long run of coated webs. This study provided the magnitude of variation arising from both down web and cross web effects. The sampled down web positions were 50 yards apart in every roll, while the sampled cross web positions were at the left, center, and right sides of the webs. These sampling points were selected for convenience and were believed to be as representative of the process as any other positions which could have been selected. There were several other possible cross web positions that could have been selected. The process capability study showed that the major source of variability was due to cross web positions. Therefore, reducing cross web variability could lead to sizable impact on reducing total variation.

In order to control mineral penetration, 8 factors were identified for the study. In addition to controlling the mineral penetration to some target, we were also interested in controlling the variability in the mineral penetration due to cross web effect. Therefore, a 28–4 fractional factorial design of 16 runs was set up to screen all the 8 factors believed to be potentially critical to controlling mineral penetration in the web process. There were no replicates or center points run at this stage in order to minimize the cost and time of running the experiment. Each condition of the experiment was run at each of the three cross web positions: left, center, and right sides of the web. The order of the experiment was completely

Table 11.9 2^{8-4} Fractional factorial design for mineral penetration

| Design identification | | | | Factors | | | | | | | | Mineral penetration | | | | |
| | | | | | | | | | | | | Cross web positions | | | Responses | |
StdOrder	RunOrder	CenterPt	Blocks	A	B	C	D	E	F	G	H	Left	Center	Right	Mean	Std dev
1	3	1	1	-1	-1	-1	-1	-1	-1	-1	-1	1.7	3.1	1.8	2.2240	0.8050
2	8	1	1	1	-1	-1	-1	1	1	1	-1	3.1	5.3	5.0	4.4440	1.2120
3	12	1	1	-1	1	-1	-1	1	1	-1	1	2.2	1.3	0.9	1.4870	0.6720
4	2	1	1	1	1	-1	-1	-1	-1	1	1	6.0	3.4	7.7	5.6850	2.1940
5	6	1	1	-1	-1	1	-1	-1	1	1	1	6.2	2.8	4.0	4.3410	1.7000
6	9	1	1	1	-1	1	-1	1	-1	1	1	2.9	2.0	1.9	2.2880	0.5530
7	1	1	1	-1	1	1	-1	1	-1	-1	-1	10.2	10.8	8.0	9.6570	1.4810
8	13	1	1	1	1	1	-1	-1	1	-1	-1	1.1	6.2	1.9	3.0830	2.7710
9	14	1	1	-1	-1	-1	1	1	1	1	1	1.9	3.2	1.2	2.0600	1.0020
10	4	1	1	1	-1	-1	1	-1	1	-1	1	2.1	0.8	1.8	1.6010	0.6960
11	16	1	1	-1	1	-1	1	-1	1	1	-1	3.0	6.4	8.0	5.8040	2.5240
12	5	1	1	1	1	-1	1	1	-1	-1	-1	1.2	5.5	4.9	3.8610	2.3160
13	11	1	1	-1	-1	1	1	-1	-1	-1	-1	2.4	1.8	2.3	2.1630	0.3340
14	10	1	1	1	-1	1	1	1	1	1	-1	5.0	7.0	4.9	5.6410	1.2100
15	15	1	1	-1	1	1	1	-1	-1	1	1	0.7	1.4	0.7	0.9160	0.3970
16	7	1	1	1	1	1	1	1	1	1	1	2.9	6.5	3.0	4.1450	2.0320

Central composite response surface design

randomized. The design matrix as well as the results obtained are provided in Table 11.9.

The analysis of the screening experiment was performed using the half normal plot in Figure 11.18. The result shows that factors G and H are statistically significant to control the mean mineral penetration. Similarly factor B is significant in controlling the variability, although the half normal plot for the standard deviation is not provided in order to minimize space. These three factors were considered for the next phase of the experimentation.

Case application of central composite design

A three-factor central composite design (Table 11.7) is a member of the most popular class of designs used for estimating the coefficients in the second order model. This design consists of 8 vertices of a 3-dimensional cube. The values of the coded factors in this factorial portion of the design are $(B, G, H) = (+1, +1, +1)$. In addition, this design consists of 6 vertices $(+1.63, 0, 0)$, $(0, +1.63, 0)$, $(0, 0, +1.63)$ of a 3-dimensional octahedron or star and six center points. If properly set up, a central composite design has the ability to possess the constant variance property of a rotatable design or may be an orthogonal design thereby allowing an independent assessment of the three factors under study. For the illustrative study, a second order response surface experiment was conducted for the three factors B, G, and H previously declared statistically significant from the above screening experiment. The design set up is provided in Table 11.10.

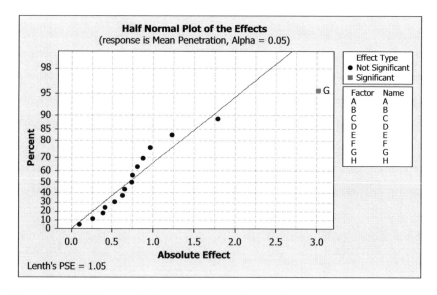

Figure 11.18 Half normal plot for moving webs application.

Table 11.10 2^3 Central composite design for mineral penetration

Central composite response surface design

| | Design identification | | | | Factors | | | Mineral penetration | | | | |
| | | | | | | | | Cross web positions | | | Responses | |
StdOrder	RunOrder	PtType	Blocks	B	G	H	Left	Center	Right	Mean	STD
1	20	1	1	−1	−1	−1	24.2	24.0	12.0	20.0667	6.9867
2	17	1	1	1	−1	−1	4.9	19.0	13.7	12.5333	7.1220
3	18	1	1	−1	1	−1	17.5	9.7	21.3	16.1667	5.9138
4	16	1	1	1	1	−1	21.7	13.3	30.3	21.7667	8.5002
5	10	1	2	−1	−1	1	19.5	25.7	45.7	30.3000	13.6923
6	14	1	2	1	−1	1	28.4	18.7	23.3	23.4667	4.8521
7	11	1	2	−1	1	1	52.4	32.0	44.7	43.0333	10.3016
8	9	1	2	1	1	1	39.5	44.7	48.3	44.1667	4.4242
9	19	0	1	0	0	0	21.2	27.3	34.7	27.7333	6.7604
10	15	0	1	0	0	0	31.4	19.0	24.3	24.9000	6.2217
11	13	0	2	0	0	0	13.6	34.0	10.3	19.3000	12.8371
12	12	0	2	0	0	0	18.6	13.7	17.0	16.4333	2.4987
13	4	0	3	0	0	0	38.6	25.7	19.7	28.0000	9.6576
14	5	0	3	0	0	0	30.4	23.0	12.0	21.8000	9.2585
15	8	−1	3	−1.633	0	0	33.4	20.7	26.7	26.9333	6.3532
16	3	−1	3	1.633	0	0	28.1	19.7	16.7	21.5000	5.9093
17	7	−1	3	0	−1.633	0	12.9	15.3	5.0	11.0667	5.3892
18	1	−1	3	0	1.633	0	31.5	37.0	31.7	33.4000	3.1193
19	6	−1	3	0	0	−1.633	13.7	13.0	13.0	13.2333	0.4041
20	2	−1	3	0	0	1.633	53.0	47.7	35.3	45.3333	9.0842

ANOVA

Tables 11.11 and 11.12 provide the corresponding ANOVA results. For the mean mineral penetration, factors G, H, H^2, and G*H interaction are statistically significant. The lack of fit (LOF) is not statistically significant ($p = 0.511$). Consequently, a second order model is fitted. The resulting response surface curve for mean penetration is provided in Figure 11.19.

Table 11.11 ANOVA for mean mineral penetration

Response surface regression: Mean versus block, B, G, H
The analysis was done using coded units.
Estimated regression coefficients for mean

Term	Coef	SE Coef	T	P
Constant	22.9601	1.2575	18.259	0.000
Block 1	2.7460	1.2412	2.212	0.058
Block 2	−2.8590	1.2412	−2.303	0.050
B	−1.2379	0.8444	−1.466	0.181
G	5.6428	0.8444	6.682	0.000
H	10.8954	1.0902	0.994	0.000
B*B	0.6572	0.8485	0.775	0.461
G*G	−0.0865	0.8485	−0.102	0.921
H*H	2.5572	0.8485	3.014	0.017
B*G	2.6375	1.0902	2.419	0.042
B*H	−0.4708	1.0902	−0.432	0.677
G*H	3.5125	1.0902	3.222	0.012

S = 3.08342, PRESS = 563.711
R-Sq = 96.11%, R-Sq(pred) = 71.19%, R-Sq(adj) = 90.77%
ANOVA for mean

Source	DF	Seq SS	Adj SS	Adj MS	F	P
Blocks	2	238.96	56.69	28.343	2.98	0.108
Regression	9	1641.34	1641.34	182.372	19.18	0.000
Linear	3	1394.66	1394.66	464.887	48.90	0.000
Square	3	90.56	90.56	30.186	3.17	0.085
Interaction	3	156.13	156.13	52.042	5.47	0.024
Residual Error 8	76.06	76.06	9.507			
Lack-of-Fit	5	48.72	48.72	9.743	1.07	0.511
Pure Error	3	27.34	27.34	9.114			
Total	19	1956.36				

The fitted response surface for the mean of the mineral penetration is obtained as

Average Penetration $= 22.96 - 1.24 * B + 5.64 * G + 10.89 * H + 0.66 * B^2$

$$-0.09 * G^2 + 2.56 * H^2 + 2.64 * B * G - 0.47 * B * H + 3.51 * G * H$$

Table 11.12 ANOVA for standard deviation mineral penetration

Response Surface Regression: STD versus Block, B, G, H
The analysis was done using coded units.
Estimated Regression Coefficients for STD

Term	Coef	SE Coef	T	P
Constant	7.77642	1.4288	5.443	0.001
Block 1	1.04467	1.4103	0.741	0.480
Block 2	−0.13647	1.4103	−0.097	0.925
B	−0.95405	0.9595	−0.994	0.349
G	−0.54150	0.9595	−0.564	0.588
H	1.77350	1.2387	1.432	0.190
B*B	0.04737	0.9641	0.049	0.962
G*G	−0.65651	0.9641	−0.681	0.515
H*H	−0.47278	0.9641	−0.490	0.637
B*G	0.67672	1.2387	0.546	0.600
B*H	−2.17992	1.2387	−1.760	0.116
G*H	−0.51550	1.2387	−0.416	0.688

S = 3.50362, PRESS = 492.447
R-Sq = 52.01%, R-Sq(pred) = 0.00%, R-Sq(adj) = 0.00%
ANOVA for STD

Source	DF	Seq SS	Adj SS	Adj MS	F	P
Blocks	2	13.110	11.419	5.710	0.47	0.644
Regression	9	93.315	93.315	10.368	0.84	0.600
Linear	3	41.209	41.209	13.736	1.12	0.397
Square	3	8.300	8.300	2.767	0.23	0.876
Interaction	3	43.806	43.806	14.602	1.19	0.373
Residual Error	8	98.203	98.203	12.275		
Lack-of-Fit	5	44.537	44.537	8.907	0.50	0.769
Pure Error	3	53.666	53.666	17.889		
Total	19	204.628				

The fitted response surface for the standard deviation is obtained as:

$$\text{Std.Dev} = 7.78 - 0.95 * B - 0.54 * G + 1.77 * H + 0.05 * B^2 - 0.66 * G^2 - 0.47 * H^2$$

$$+0.68 * B * G - 2.18 * B * H - 0.52 * G * H$$

The standard deviation is modeled rather than the commonly used log s^2 because there are three cross web positions (m = 3) under study. The theory for log s^2 transformation requires a moderate to large amount of replication (m > 10). For small m, we prefer to model the standard deviation, which is the square root transformation. In addition, using the

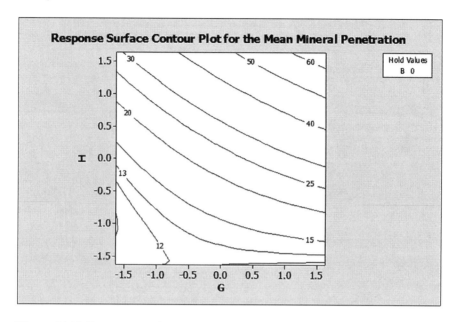

Figure 11.19 Response surface plot for the mean mineral penetration.

standard deviation will make it easier for engineers to correctly interpret the results from the model. In general, one would expect the true order of the variance model to be lower than the order of the model of the mean. However, since we run a design appropriate for a full second order model, we would fit a second order for the standard deviation. The response surface curve obtained for standard deviation is provided in Figure 11.20 for the situation when factor B is set at the center. In this case, the minimum variability occurs when factor G is at its highest level and factor H is at the lowest level.

Response surface optimization

The goal of this experiment is to find the conditions which minimize the cross web variability while achieving a specification range of 15 to 20 for the mean mineral penetration. This therefore suggests using the standard deviation as the primary response and the mean as the secondary response. We constrain our variability between 0 and 5 for our optimization search. The result obtained is provided in Figure 11.21. The optimum operating windows are clearly provided in Figures 11.21 through 11.23. These results suggest that when B is operating at the center level, factors H should be set at its lowest level and factor G should be set at its highest level. Similarly, when factor B is set at low, then factor H should be set

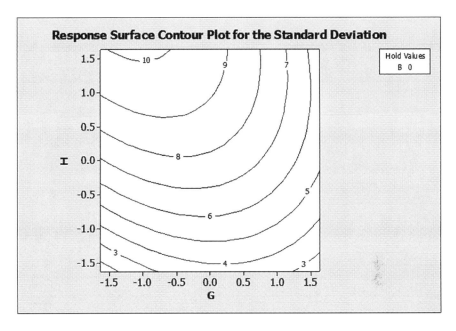

Figure 11.20 Response surface plot for mineral penetration standard deviation.

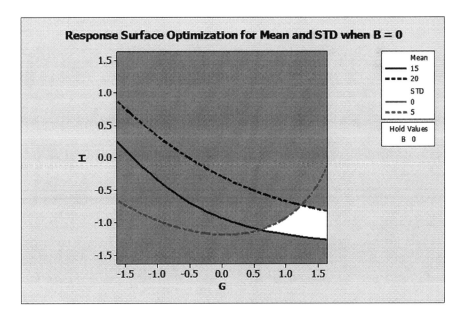

Figure 11.21 Factor B is at center level while both G and H are varied.

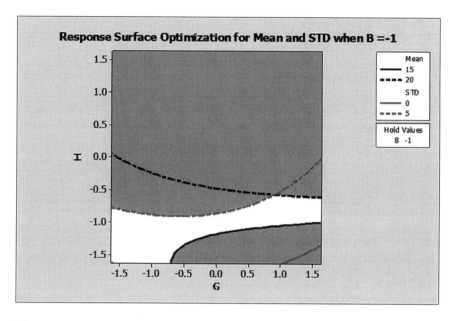

Figure 11.22 Factor B is set at low level while both factors G and H are varied.

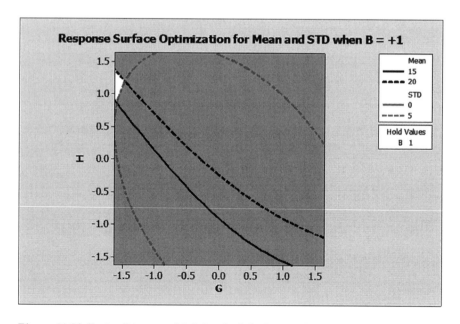

Figure 11.23 Factor B is set at high level while factors G and H are varied.

at its lowest level, while factor G should be at its lowest or center level. Response surface optimization plots are provided for the average and standard deviation.

This study has significant impact on the way we control cross web variability. Several optimal (Figures 11.21 through 11.23) operating conditions with similar desirable properties are available to engineers for controlling the cross web variability as well as achieving the mineral penetration target. In addition, the advantage of being able to plot the desirability surfaces for both mean and standard deviation to determine their sensitivities to small changes in the levels of the control factors is significant for future control of cross web variability. This dual response optimization approach is applicable to multiple responses as well, and it is consistent with the positive aspects of the Taguchi contributions.

References

Box G.E.P. (1985). Discussion of off-line quality control, parameter design, and the Taguchi methods, *Journal of Quality Technology*, 17, 198–206.

Box G.E.P., Hunter W.G., and Hunter J.S. (1978). *Statistics for Experimenters*, John Wiley & Sons, New York.

Myers, R.H. and Carter, W.H. (1973). Response surface techniques for dual response systems, *Technometrics*, 15, 301–317.

Taguchi, G. (1986). *Introduction to Quality Engineering: Designing Quality into Products and Processes*, Kraus International Publications, White Plains, NY.

Taguchi, G. and Wu, Y. (1985). *Introduction to Off-line Quality Control*, Central Japan Quality Control Association, Nagaya, Japan.

Vining, G. G and Myers, R.H, (1990). Combining Taguchi and response surface philosophies: A dual response approach, *Journal of Quality Technology*, 22, 38–45.

Bibliography

Ayeni, B.J. (1991). *Design Resolution. Statistically Speaking*. 3M Internal Publication, 3M IS&DP Statistical Consulting, August.

Ayeni, B.J. (1994). "Using response surface design to achieve target with minimum variance," *Proceedings of the 2nd Africa–USA International Conference on Manufacturing Technology*, August 1994, Pages 1–10.

Ayeni, B.J. (1999). *Application of Dual-Response Optimization for Moving Web Processes*, 3M Internal Publication, IT Statistical Consulting. 3M Company, St. Paul, MN.

Badiru, A.B. and B.J. Ayeni (1993). *Practitioner's Guide to Quality and Process Improvement*, Chapman & Hall, London.

Box G.E.P. and N.R. Draper (1987). *Empirical Model Building and Response Surfaces*, John Wiley and Sons, New York.

Hicks, C.R. (1982). *Fundamental Concepts in the Design of Experiments*, 3rd edition, Holt, Reinhart, and Winston, New York.

Myers, R.H., Khuri, A.I., and Vining, G. (1992). Response surface alternatives to the Taguchi robust parameter design approach, *The American Statistician*, 46(2), 131–139.

Yates, F., (1935). Complex experiments. *Supplement to the Journal of the Royal Statistical Society*, II(2), 181–247.

chapter twelve

Optimal experimental designs

Introduction

In recent years, it has become popular in manufacturing industries to use fractional factorial designs, as presented in previous chapters, to study a large number of factors for industrial experimentation so as to reduce variability, cost, time, and to optimize the manufacturing processes. In this section we implement a general algorithmic approach (Li, Li, and Ayeni, 1999) for the construction of optimal balanced experimental designs. It is to be noted that many existing optimal designs are not balanced, and the unbalanced nature of these designs can be disadvantageous in practice, leading to an incorrect estimate of effects due to collinearity between factors or confounding issues. In order to address these issues, we recommend the use of column-wise-pairwise algorithm (CP algorithm) to generate optimal designs. Li (1997) proposed a new algorithm for the construction of optimal designs while retaining the small number of runs as well as the balance of factors (see also Li and Wu, 1997). The two novel ideas in the algorithm are to interchange columns instead of rows and to exchange pairs instead of searching over all candidates. Thus, it is called *CP algorithm*.

There are several areas to which the CP algorithms can be applied. Some types of problems include supersaturated design, mixed-level design, and many other model-driven designs in which orthogonality cannot be retained. The proposed algorithmic procedure can spread out the non-orthogonal pairs of designs more evenly, and consequently, lead to designs with higher efficiency. The CP algorithms can also be applied to design repair problems, in which a submatrix of the design matrix is optimized.

In industrial experimentations, experimental designs are frequently constructed to estimate all main effects and a small number of pre-specified interactions. The requirement that the experimenter knows which interactions are likely to be active in advance is a major limitation of this approach. Therefore, a method to construct designs for estimation of main effects and *any* combinations of q interactions, where q is specified by the user, is desired. This is viewed as in issue of model robust design. A modified CP algorithm called restricted pairwise exchange algorithm is used to address this problem.

A Java applet named *GO!* is written to implement this general algorithmic approach of the construction of optimal balanced designs. A
number of examples are used to explain how to use *GO!*. All examples
are drawn from industry-related areas.

CP algorithm

A CP algorithm is an effective algorithm to construct optimal designs.
The idea of the CP algorithm is to change columns, instead of rows, in
the iterative procedures. And when a column is to be exchanged, instead
of searching over the set of all candidate vectors, a subset of pairs of elements is considered; thereafter, the number of computations can be substantially reduced. Here is the description of the algorithm:

> *Step 1.* Randomly choose a starting design.
>
> *Step 2.* For each column, choose the best pair to exchange that gives the
> highest *criterion*. The design matrix is updated accordingly.
>
> *Step 3.* Repeat Step 2 if an improvement is made. Otherwise the result-
> ing design is kept in the pool of "good designs." Step 1–3 is called
> one *trial*.
>
> *Step 4.* Perform trials repeatedly (i.e., repeat Step 1–3 with different
> starting designs) and pick the best design from the pool of "good
> designs."

There are different criterions to compare and evaluate designs. In this
project, the normalized D criterion is chosen because of its many good
properties. The normalized D is defined as $|\mathbf{X}^T\mathbf{X}|^{1/f}$, where $\mathbf{X}$ is the design
matrix and f is the number of columns of $\mathbf{X}$.

Browser requirement:

GO! is a Java applet written and compiled using JDK1.2.1. It can be run on
both *Internet Explorer* and *Netscape Navigator*. The browser is required to
support Java™ Swing. If your browser is not Swing-ready, you can down-
load Java™ Runtime Environment Version 1.2.2.

Run GO!:

There are three pages in *GO!*. To switch among the three pages, click the
choice box (see Figure 12.1) at the top of each page and then select the
page you want to go. Before design searching, you need to first set up the
Parameter Page. In some cases, you may want to input the starting matrix
by yourself, rather than generating it randomly. Then you need to go to
the *Matrix Page* to input the whole starting matrix or part of the matrix.
After finishing the parameters setting and the matrix input (if needed),
go to the *Display Page*. There you will see the display of your parameter

Parameter Page	▼
Parameter Page	
Matrix Page	
Display Page	

Number of Interact		Number of Trials	
Fixed Row		Fixed Column	
List of Factor Levels			
List of Interactions			

Figure 12.1 Choice box.

setting, as a double check to assure right input. Then click the "*GO!*" button at the bottom of the Display Page. After a few seconds, the resulting matrix will be obtained and displayed.

Parameter page

Here is a look at the Parameter Page (see Figure 12.2), followed by the explanations of each field in this page. More explanations with examples can be found in the *Applications and Demos* section.

Number of Runs: number of runs to perform in the experiment. This field must be filled before running *GO!* A warning will be shown in the Display Page if this field is blank and the program will stop.

Number of Factors: number of factors in the experiment. This field must be filled before running *GO!* A warning will be shown in the Display Page if this field is blank and the program will stop.

Number of Interact: number of interactions in the experiment. This field can be left blank if there are no interactions. If it is blank, the default value is 0.

Number of Trials: the number of trials to perform to find the optimal design. The search starting from one starting matrix is called one trial. (see also *CP Algorithm* section) If the design is complicated, the search from more than one starting matrix is expected. The usual number of trials is 5 or 10. You can also leave this field blank, with the default value as 1.

Fixed Row: In some cases, the starting matrix may have a fixed submatrix and such fixed submatrix will be input by user. The number of rows of the submatrix is specified in this field. If this field is left

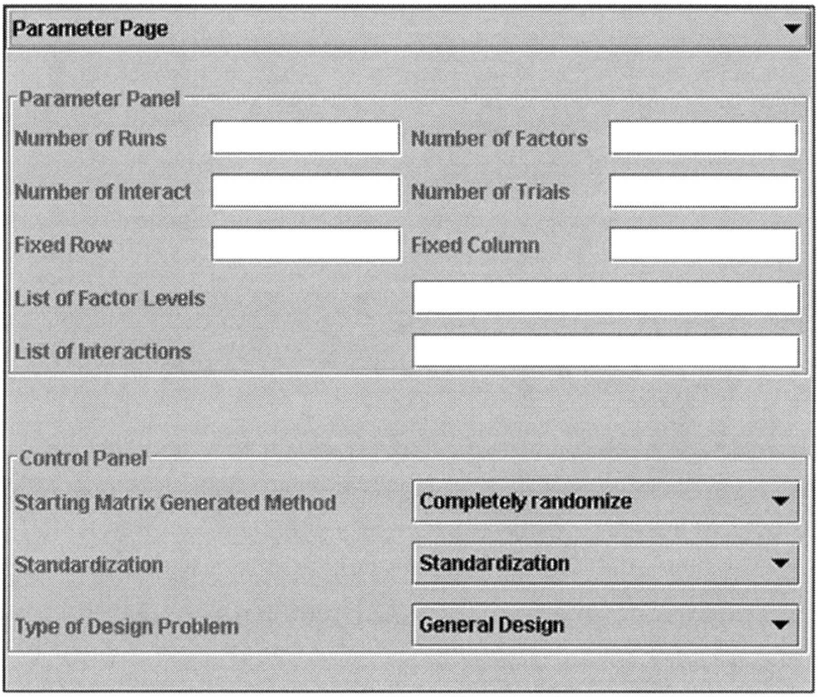

Figure 12.2 Parameter page.

blank, then the default value is set as 0, meaning that there is no fixed submatrix in the starting matrix.

Warning: It is required to put the fixed submatrix, if there is any, at the upper left corner of the full matrix.

Fixed Column: The number of columns of the submatrix is specified in this field.

List of Factor Levels: the level of each factor is specified in this field. For example, if an experiment has four 2-level factors, the list "2 2 2 2" needs to be filled in this field. A blank space is between each level number. This field must be filled before running *GO!*. A warning will be shown in the Display Page if this field is blank and the program will stop.

List of Interactions: If the number of interactions is more than 0, this field needs to be filled to specify the interactions. (There is an exception to this rule in the Model Robust Design problem. See *Applications and Demos* section for more information.) For example, if an experiment has two interactions x_1*x_1, x_2*x_2, the list "1*1 2*2" needs to be filled in this field. A star "*" is between each interaction item and a blank space is between different interactions.

Starting Matrix Generated Method: There are three types of starting matrix generated methods. The default one is *Completely randomize*, which means the starting matrix will be generated completely randomly. If there is a fixed submatrix in the starting matrix, then the **Fixed Row** and **Fixed Column** need to be specified and the *Read Fixed Matrix Only* should be chosen. Sometimes the starting matrix is wholly fixed, e.g., we want to verify the criterion of an obtained matrix, then we should choose *Read from Matrix Page* to get the input of starting matrix from the Matrix Page.

Standardization: There are two choices, "Standardization" and "No standardization". The default selection is "Standardization". A matrix will be standardized if "Standardization" is chosen and the best criterion could be reached is 1.

Type of Design Problem: There are different types of design problems. You can specify the type or, in most cases, you can use *General Design*, which is the default selection. There are three types of design problems in which you *cannot* use *General Design*. The first one is *Central Composite Design-blocking*, the second one is *Design Repair-row permutation*, and the third is *Model Robust Design*. More explanations about these types can be found in *Applications and Demos* section.

Matrix page

The Matrix Page (see Figure 12.3) is to accept user input of the whole or part of the starting matrix. After finishing matrix input, click the "Done" button at the bottom of the page. Then the input data will be read and stored. If any table cell which should be filled with data is left blank, a warning will appear in the Display Page and the program will stop.

Warning: If a factor has more than 2 levels, then the input of the factor should only have numbers like 1, 2, 3, etc. For example, for a 3-level factor, the input should only contain 1, 2, and 3. If it looks like –1, 1, 2 or –1, 1, 3, an error will happen. Note that this rule doesn't apply to 2-level factor. So you can input both 1, 2 and 1, –1.

Warning: Since the table data update is focus-sensitive, i.e., only when one table cell loses the focus will the data of this cell be updated, please leave the last cell of input before clicking the "Done" button. Otherwise the last input cell data will not be read.

Display page

In Display Page (see Figure 12.4), you can see all the output information including the warning, the searching procedures, and the obtained optimal design matrix. After parameter setting and matrix input (if needed), click the *"GO!"* button. If there is any must-be-filled field left blank or

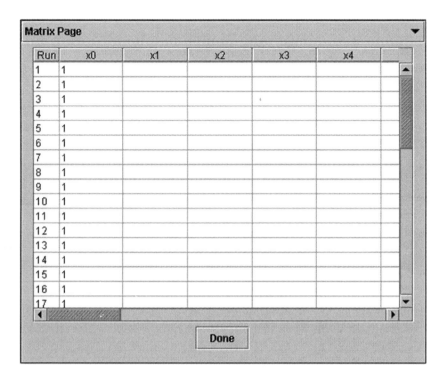

Figure 12.3 Matrix page.

invalid, a corresponding warning will be shown and the program will stop. Go back to reset the missing value(s) then click the "*GO!*" button again to run the program.

Copy the obtained matrix
After getting the optimal matrix, it will be displayed in the Display Page. To conduct further analysis, choose the output matrix and use *Ctrl-C* to copy it to the system clipboard. Then you can either paste it to a file or to the table of any statistical software, like *Minitab*.

Invalid input
If there is any invalid input, e.g., a field which should accept an integer gets a character input, an error message will appear. (See Figure 12.5) Double-click the "OK" button in the error message window and re-input the valid data.

Update the applet
If you want to start with another completely new design, click the "Refresh" button on the menu bar in *Internet Explorer*, or, the "Reload"

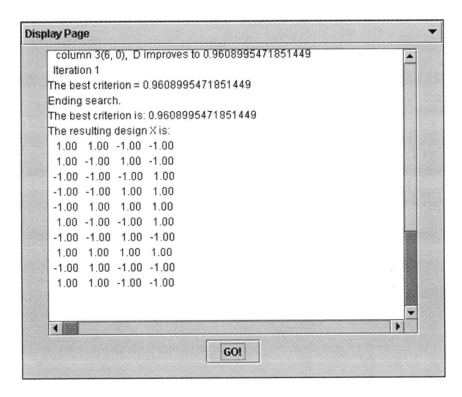

Figure 12.4 Display page.

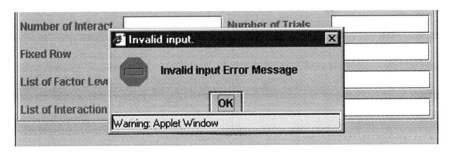

Figure 12.5 Invalid input error message.

button on the menu bar in *Netscape Navigator* to update the applet. In such cases, an error might happen if the applet is *not* updated.

Example 1: Non-orthogonal design
Description
Ten runs are allowable for an experiment. How do we construct a 10-run two-level design to study four factors? (Li, 1997)

Number of Runs	10	Number of Factors	4
Number of Interact		Number of Trials	
Fixed Row		Fixed Column	
List of Factor Levels		2 2 2 2	
List of Interactions			
Starting matrix generated method		Complete randomize	
Standardization		Standardization	
Type of Design Problem		General Design	

Figure 12.6 Entry of requirements for design construction.

Since orthogonal designs require the run size to be a power of four, the purpose of this example is to show how the CP algorithm could be used for constructing a general non-orthogonal design.

Setting of *GO!* (Figure 12.6):

Explanation

This is a simple example of using the *GO!*. We need to fill three must-be-filled fields, which are "Number of Runs," "Number of Factors," and "List of Factor Levels." The starting matrix generated method is "Completely randomize." You can choose "General Design" as the design type.

Result

The resulting design has $D = 0.961$. The correlation of each pair of factors is 0.2 (Table 12.1).

Table 12.1 A 10-run two-level design to study 4 factors

Run	A	B	C	D
1	1	−1	1	1
2	−1	1	−1	−1
3	−1	1	1	1
4	−1	−1	−1	1
5	−1	−1	1	−1
6	1	1	1	−1
7	1	1	−1	1
8	1	−1	−1	−1
9	−1	1	−1	1
10	1	−1	1	−1

Correlation: A, B, C, D

	A	B	C
B	−0.200		
	0.555		
C	0.200	−0.200	
	0.555	0.555	
D	−0.200	0.200	−0.200
	0.555	0.555	0.555

Cell Contents: Pearson correlation
    P-Value (Table 12.2)

This design is represented geometrically as a cube in Figure 12.7.

Analysis of effects of elastics retraction using coded form (Figure 12.8)

Regression equation in coded units

$$\text{Retraction at } 75\% = 0.3536 + 0.002138\,A - 0.04739\,B - 0.05343\,C$$

$$+0.002306\,D - 0.006674\,A * B$$

$$+0.009007\,A * C + 0.1118\,A * D$$

$$+0.05188\,B * C - 0.000725\,A * B * C$$

Table 12.2 Analysis of effects using coded form: Elastics retraction at 75%

Run	A	B	C	D	Retraction at 75%
1	1	−1	1	1	0.428
2	−1	1	−1	−1	0.430
3	−1	1	1	1	0.191
4	−1	−1	−1	1	0.398
5	−1	−1	1	−1	0.387
6	1	1	1	−1	0.194
7	1	1	−1	1	0.409
8	1	−1	−1	−1	0.391
9	−1	1	−1	1	0.211
10	1	−1	1	−1	0.200

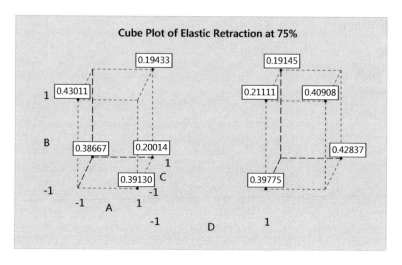

Figure 12.7 Cube plot of elastic retraction at 75%.

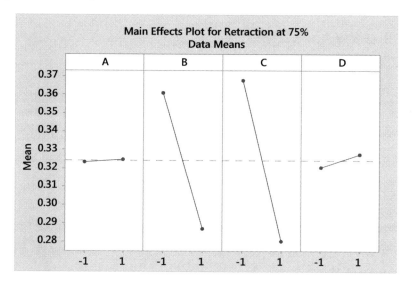

Figure 12.8 Main effects plot for retraction at 75%.

From the main effects plot above, we can see that the main effects of factors B and C have the largest changes among the four factors, and therefore, are statistically significant in controlling the elastics retraction. The statistically significant effect of factors B and C are also confirmed by the Pareto chart of the effect plot (Figure 12.9).

The regression model equation in coded form is immediately provided below the main effects plot.

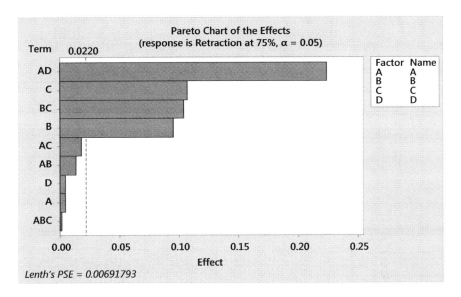

Figure 12.9 Pareto chart of the effect.

The interaction plot as well as the Pareto chart of effects plot showed that factors A and D interactions are statistically significant. In addition, factors B and C interactions also have statistically significant effects. For example, the interaction plot of factors A and D in Figure 12.10 shows that at each level of factor D, the effect of factor A is crosses the effect of factor D indicating that there is strong interaction effect between factors A and D. This means that factor D at high level will consistently provide high level of elastics retraction of about 0.425 at high level of factor A but at a low level of factor A the elastics retraction is significantly reduced to a low of about 0.260 (Figure 12.11).

Example 2: Model-driven design

Description

An experiment is designed to investigate four 2-level factors: 1, 2, 3, and 4. From the prior experience, the interactions 1*2 and 3*4 may be significant and should not be aliased with the main effects. If the budget only allows eight runs to be conducted, how do we construct a design with high efficiency? (Li, 1997)

Setting of *GO!* (Figure 12.12)

Explanation

This is an example of a design with possible interactions that need to be estimated clean, clear of main effects. In addition to three must-be-filled fields in the design generation form of *GO!*, which are "Number of Runs," "Number of Factors," and" List of Factor Levels," there are

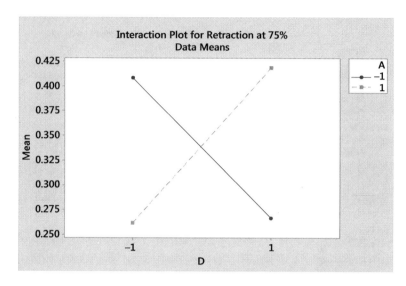

Figure 12.10 Interaction Plot for Elastic Retraction at 75% with Factor D at the Horizontal.

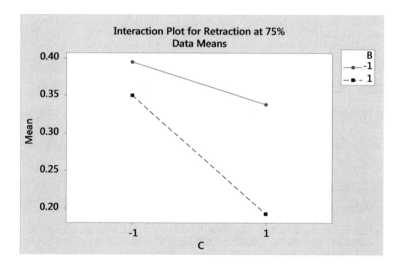

Figure 12.11 Interaction plot for elastic retraction at 75% with factor C at the horizontal.

two interactions – 1*2, 3*4 – in the design form that must be entered. So "Number of Interact" is set as 2 and "List of Interactions" is set as "1*2 3*4." The starting matrix generated method is "Completely randomize." You can choose either "Model-driven Design" or "General Design" as the design type.

Number of Runs	8	Number of Factors	4
Number of Interact	2	Number of Trials	
Fixed Row		Fixed Column	
List of Factor Levels		2 2 2 2	
List of Interactions		1*2 3*4	
Starting matrix generated method		Completely randomize	
Standardization		Standardization	
Type of Design Problem		Model-driven Design / General Design	

Figure 12.12 Construction of 8 runs of four 2-level factors with interactions.

Result

The resulting design has $D = 0.9999$. The correlation of each pair of factors is 0.0 (Table 12.3).

Correlation: A, B, C, D

	A	B	C
B	0.000		
	1.000		
C	0.000	0.000	
	1.000	1.000	
D	0.000	0.000	0.000
	1.000	1.000	1.000

Cell Contents: Pearson correlation
P-Value

The design in Table 12.3 was used to study the effects of four factors on substrate temperature of a polycarbonate material. The experiment was run in a randomized order and the results obtained are provided in Table 12.3. Due to the cost constraint, we can only run 8 experiments and no replicates or center points were run for the same cost reason. The replicates could have enabled us to obtain an estimate for pure experimental error as well as ANOVA table with the corresponding p-values. However, because of the above restrictions, we performed the analysis of the results on Table 12.4 using the Pareto chart of effect plots in Figure 12.14 for interpretation.

From the main effect plot below, we can see that only factor D is statistically significant in controlling the substrate temperature of polycarbonate material under study. The statistical significant effect of factor D

Table 12.3 Design matrix for 8 runs with 4 factors

Run	A	B	C	D
1	−1	1	−1	1
2	1	1	1	−1
3	−1	−1	1	−1
4	1	−1	1	1
5	−1	1	1	1
6	1	−1	−1	1
7	1	1	−1	−1
8	−1	−1	−1	−1

Table 12.4 Analysis of effects using coded form

	Coded Form				Un-coded Form				Substrate
Run	A	B	C	D	A	B	C	D	Temp deg F
1	−1	1	−1	1	Material 1	50	75	270	223
2	1	1	1	−1	Material 2	50	500	110	125
3	−1	−1	1	−1	Material 1	5	500	110	105
4	1	−1	1	1	Material 2	5	500	270	268
5	−1	1	1	1	Material 1	50	500	270	240
6	1	−1	−1	1	Materia 2	5	75	270	250
7	1	1	−1	−1	Material 2	50	75	110	99.5
8	−1	−1	−1	−1	Material 1	5	75	110	98

is also confirmed by the Pareto chart of effect plot. The regression model equation in coded form is immediately provided below the main effects plot (Figure 12.13).

Regression equation in coded form

$$\text{Substrate Temp deg F} = 176.1 + 9.563\,A - 4.188\,B + 8.437\,C + 69.19\,D$$

$$+ 2.438\,A * C + 2.188\,B * C + 0.3125\,C * D$$

The Pareto chart of effects plot (Figure 12.14) as well as the interaction plot (Figure 12.15) showed that there is no statistically significant interaction effect. For example, the interaction plot of factors C and D below shows that at each level of factor D, the effect of factor C is parallel to the effect of factor D indicating that there is no interaction effect between factors C and D. This means that factor D at high level will consistently provide high level of substrate temperature. If we want to control the

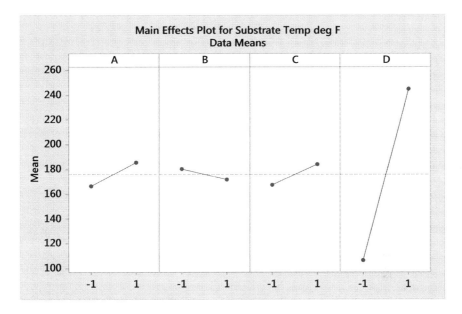

Figure 12.13 Main effect plot for substrate temp deg F.

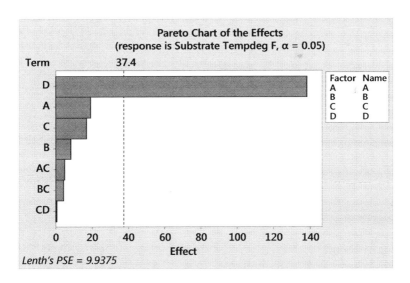

Figure 12.14 Pareto chart of the effects.

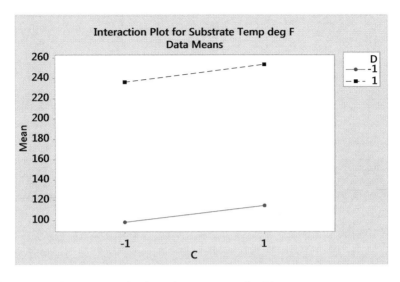

Figure 12.15 Interaction plot for substrate temp deg F.

polycarbonate material in order to achieve a low level of substrate temperature, we will have to put factor D at low level in order to achieve this goal (Table 12.5).

From the main effects plot in Figure 12.16, we can see that only factor D is statistically significant in controlling the substrate temperature of polycarbonate material under study. The statistical significant effect of factor D is also confirmed by the Pareto chart of the effect plot (Figure 12.16).

Table 12.5 Analysis of effects using un-coded form: Substrate temperature deg F

	Coded Form				Un-coded Form				
Run	A	B	C	D	A	B	C	D	Substrate Temp deg F
1	−1	1	−1	1	Material 1	50	75	270	223
2	1	1	1	−1	Material 2	50	500	110	125
3	−1	−1	1	−1	Material 1	5	500	110	105
4	1	−1	1	1	Material 2	5	500	270	268
5	−1	1	1	1	Material 1	50	500	270	240
6	1	−1	−1	1	Materia 2	5	75	270	250
7	1	1	−1	−1	Material 2	50	75	110	99.5
8	−1	−1	−1	−1	Material 1	5	75	110	98

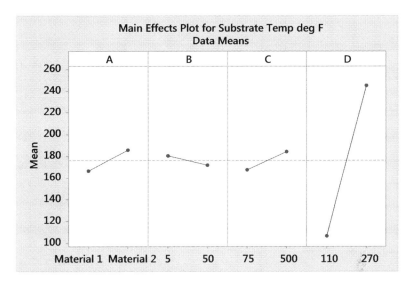

Figure 12.16 Main effects plot for substrate temp deg F.

Regression equation in un-coded units

$$\text{Substrate Temp deg F} = 10.07 + 6.265\,A - 0.3176\,B + 0.02363\,C + 0.8596\,D$$

$$+ 0.01147\,A * C + 0.000458\,B * C + 0.000018\,C * D$$

The Pareto chart of effects plot (Figure 12.17) as well as the interaction plot (Figure 12.18) showed that there is no statistically significant interaction effect. For example, the interaction plot of factors C and D below shows that at each level of factor D, the effect of factor C is parallel to the effect of factor D indicating that there is no interaction effect between factors C and D. This means that factor D at high level of 270 will consistently provide high level of substrate temperature from a low of 240° F to a high of 250° F substrate temperatures. However, if we want to control the polycarbonate material in order to achieve a low level of substrate temperature, we will have to put factor D at low level of 110 in order to achieve this goal which provided us with c as 100° F and as high 110° F.

Example 3: Mixed-level design
Description
To reduce the geometric distortion of critical part characteristics of a rear axle gear, the heat-treatment process is identified as the primary target for improvement. Seven factors are being considered. Factor A has 3 levels corresponding to the three sources of the gear. Each of the remaining

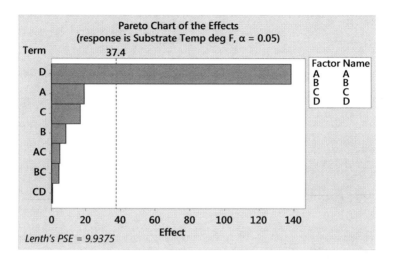

Figure 12.17 Pareto chart of the effects.

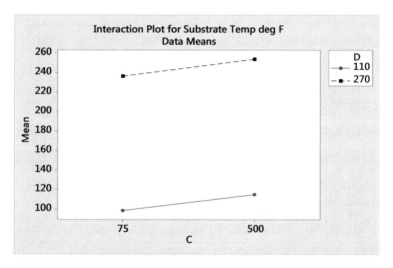

Figure 12.18 Interaction plot for substrate temp deg F.

factors (temperature and time in furnace, quench-oil type, temperature, furnace cycle time, and operating mode) has 2 levels. How do we construct a 12-run array to serve this purpose? (Li, 1997)

Setting of *GO!*(Figure 12.19)

Explanation

This is an example of mixed-level design. Based on our previous experience, there are *local* optimal designs for this problem; therefore, we set

Number of Runs	12	Number of Factors	7
Number of Interact		Number of Trials	5
Fixed Row		Fixed Column	
List of Factor Levels		3 2 2 2 2 2 2	
List of Interactions			
Starting matrix generated method		Completely randomize	
Standardization		Standardization	
Type of Design Problem		Mixed-level Design	

Figure 12.19 Construction of a 12-run mixed level design with 7 factors.

"Number of Trials" at five to get as good a design as possible. The starting matrix generated method is "Completely randomize."

   Result
Here is a resulting design. This design has *D* equal to 0.974 and only 2 non-orthogonal pairs (Table 12.6).

Analysis of effects using un-coded form: percentage tensile extension at yield

The design in Table 12.6 was used to study the effects of seven factors on the tensile extension of an elastic product during a new product development activity. The tensile extension at yield is a major performance property of the new product and the new product development team would like to know which of the seven control factors are critical in controlling

Table 12.6 Design matrix for a 12-run mixed level design with 7 factors (Coded)

Run	A	B	C	D	E	F	G
1	2	−1	−1	1	−1	1	1
2	1	−1	1	1	−1	−1	−1
3	3	−1	1	1	−1	−1	1
4	1	1	−1	1	1	1	−1
5	3	−1	−1	−1	1	−1	−1
6	2	1	1	−1	1	−1	1
7	1	1	−1	−1	−1	−1	1
8	3	1	−1	1	1	1	1
9	3	1	1	−1	−1	1	−1
10	2	1	1	1	1	−1	−1
11	1	−1	1	−1	1	1	1
12	2	−1	−1	−1	−1	1	−1

the key product performance property which is the tensile extension at yield (Table 12.7).

Regression equation in un-coded units

$$\text{Tensile Ext at Yield}\,(\%) = 1031 - 20.54\,A - 4.009\,B - 0.8491\,C - 58.58\,D$$

$$+9.055\,E - 0.009729\,F - 6.963\,G + 0.07840\,A * B$$

$$+1.684\,A * D + 0.2991\,A * G + 0.007430\,B * C$$

```
Coded coefficients

Term            Effect      Coef

Constant                    300.10
A              -33.190      -16.59
B             -120.89       -60.44
C              -12.889       -6.45
D              -24.720      -12.36
E               18.111        9.06
F               -0.389       -0.19
G               10.278        5.14
A*B             35.280       17.64
A*D             12.629        6.31
A*G             29.910       14.95
B*C             30.090       15.05
```

Table 12.7 Design matrix for a 12-run mixed level design with 7 factors (Un-coded)

Run	A	B	C	D	E	F	G	Tensile Ext at Yield (%)
1	25	50	50	3	1	90	30	366
2	20	50	140	3	1	50	10	356
3	30	50	140	3	1	50	30	310
4	20	140	50	3	3	90	10	230
5	30	50	50	1.5	3	50	10	343
6	25	140	140	1.5	3	50	30	275
7	20	140	50	1.5	1	50	30	230
8	30	140	50	3	3	90	30	255
9	30	140	140	1.5	1	90	10	226
10	25	140	140	3	3	50	10	240
11	20	50	140	1.5	3	90	30	391
12	25	50	50	1.5	1	90	10	380

The main effect plot (Figure 12.20) as well as the Pareto chart of effects (Figure 12.21) showed that the most critical factor for controlling the tensile extension at yield is factor B. It can be seen that factor B has the largest absolute change of about 120.89% from a low level of 50 to a high of 140. This means that there is a significant drop of about 120.89% of tensile extension at yield when you change factor B from a low of 50 to a high of

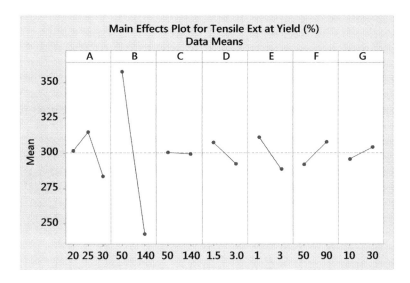

Figure 12.20 Main effects plot for tensile ext at yield.

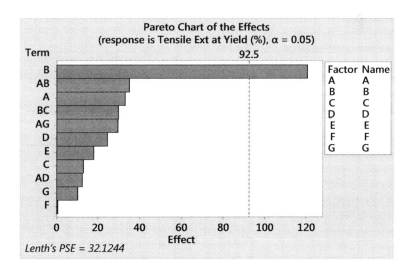

Figure 12.21 Pareto chart of effects.

140. Factor B therefore contributed the largest effect on the tensile extension of the elastics product and served as our major controller of the product. The main effect changes from a low level to the high level for each of the other factors are statistically insignificant and, therefore, are not important by themselves for controlling the tensile extension at yield. In addition, from the Pareto chart, we found that there are no significant interaction effects.

Example 4: Design repair with row permutations
Description
To reduce the master transfer function variability of Mass Air Flow Sensors (MAFS) in cars, an experiment is conducted. Eight factors are chosen to be investigated. x_1: element position – vertical; x_2: element position – horizontal; x_3: element position – at an angle; x_4: operating temperature; x_5: bypass interior roughness; x_6: bypass back plate leaks; x_7: bypass tabs off-center; x_8: bypass inlet tip radius/angle.

The original design is a 16-run experiment with the defining relationships: 5=234, 6=134, 7=123, and 8=124. The sensing elements are produced in a factory in Japan, and the settings of x_1 and x_3 are varied according to columns 1 to 3 in Table 12.8. At the next stage of the experiment, they are located within the bypass whose conditions are varied according to columns 4 to 8 of the same table. After the elements are produced, however, it is found that the values of the 1st and 9th element of column 3 are about

Table 12.8 Design matrix for a 16-run experiment with 8 factors

Run	x_1	x_2	x_3	x_4	x_5	x_6	x_7	x_8
1	−1	−1	−1 (3)	−1	−1	−1	−1	−1
2	1	−1	−1	−1	−1	1	1	1
3	−1	1	−1	−1	1	−1	1	1
4	1	1	−1	−1	1	1	−1	−1
5	−1	−1	1	1	1	1	1	−1
6	1	−1	1	1	1	−1	−1	1
7	−1	1	1	1	−1	1	−1	1
8	1	1	1	1	−1	−1	1	−1
9	−1	−1	−1 (3)	−1	1	1	−1	1
10	1	−1	−1	−1	1	−1	1	−1
11	−1	1	−1	−1	−1	1	1	−1
12	1	1	−1	−1	−1	−1	−1	1
13	−1	−1	1	1	−1	−1	1	1
14	1	−1	1	1	−1	1	−1	−1
15	−1	1	1	1	1	−1	−1	−1
16	1	1	1	1	1	1	1	1

5° C, not –5° C as planned. Thus, the design matrix is no longer orthogonal. Since re-producing two flawed sensors are expensive and time-consuming, how do we adjust the submatrix of columns 4 to 8 such that the overall design is as orthogonal as possible? (Li, 1997).

In Table 12.8, factor 3, element position – at an angle, has two levels. –1 means –5°C and 1 means 0° C. When 5° C appears in the real experiment result, it will be represented as 3 in the table and factor 3 will be treated as a quantitative factor.

The main challenge here is that the submatrix containing columns 1, 2, and 3 cannot be changed. Denote the design matrix by $X = (X_1, X_2)$, where $X_1 = (1, 2, 3)$. Then the objective is to find an optimal matrix X_2 such that the overall design X is as good as possible. The engineering collaborator further suggests that the planned combinations represented by the rows of X_2 cannot be changed. So, the adjustment can only be made by the permutations on these rows. Such a problem is called design repair with row permutations.

Setting of *GO!* (Figure 12.22)

Explanation

This is an example of a design repair problem. Since the submatrix X_1 is fixed, we need to set "Fixed Row" as 16, "Fixed Column" as 3. Our starting matrix will look like the data in Table 12.8, with *–1* replaced by *3* on the 1st and 9th row of factor 3. Therefore, the level of factor 3 is 1, instead of 2 now. Also, we *should* choose "Read from Matrix Page" as the starting matrix generated method. Because of the unchanged combinations represented by the rows of X_2, we *should* use "Design Repair-row permutation" as the type of design.

Result

The starting design has D criterion equal to 0.928. There are 3 non-orthogonal pairs in this design. After exchanging row 1 and 5, and row 9 and 13,

Number of Runs	16	Number of Factors	8
Number of Interact		Number of Trials	
Fixed Row	16	Fixed Column	3
List of Factor Levels		2 2 1 2 2 2 2 2	
List of Interactions			
Starting matrix generated method		Read from MatrixPage	
Standardization		Standardization	
Type of Design Problem		Design Repair -row permutation	

Figure 12.22 Construction of a 16-run fixed row and fixed column 8 factors design.

the D criterion increases to 0.949 and the design only has 2 non-orthogonal pairs.

Example 5: Central composite design with blocking
Description

There are three 2-level factors, x_1, x_2, and z (a blocking factor) in a 10-run experiment. The x's will form a rotatable central composite design and z has two choices 1 and −1. The design structure is shown in Table 12.9. How do we choose the levels of z for the ten runs to get an efficient design? (Wu and Ding, 1998) (Table 12.9 and Table 12.10).

Setting of *GO!* (Figure 12.23)

Table 12.9 Design matrix for a 10-run 3-factor central composite design

Run	x_1	x_2	z	$x_1{}^*x_1$	$x_1{}^*x_2$	$x_2{}^*x_2$	$x_1{}^*z$	$x_2{}^*z$
1	0	0	1					
2	0	0	1					
3	1.414	0	1					
4	−1.414	0	1					
5	0	1.414	1					
6	0	−1.414	−1					
7	1	1	−1					
8	−1	1	−1					
9	1	−1	−1					
10	−1	−1	−1					

Explanation

This example illustrates how to use a central composite design with a blocking factor. For the starting matrix, the first 6 rows and 2 columns are always fixed. Therefore, we set "Fixed Row" as 6 and "Fixed Column" as 2. For runs 7–10 of factors x_1 and x_2, you can either generate it randomly or fix it according to a standard fractional factorial design. If you choose to fix those runs, then you need to set "Fixed Row" as 10 instead of 6. In this case, the starting matrix generated method should be "Read fixed matrix only." The type of design problem should be "Central Composite Design-blocking." We *cannot* use "General Design" here.

Table 12.10 Design matrix of a 10-run 3-factor central composite design

Run	x_1	x_2	z	$x_1{}^*x_1$	$x_1{}^*x_2$	$x_2{}^*x_2$	$x_1{}^*z$	$x_2{}^*z$
1	0	0	1	0.000	0.000	0.000	0.000	0.000
2	0	0	1	0.000	0.000	0.000	0.000	0.000
3	1.414	0	1	1.999	0.000	0.000	1.414	0.000
4	−1.414	0	1	1.999	0.000	0.000	−1.414	0.000
5	0	1.414	1	0.000	0.000	1.999	0.000	1.414
6	0	−1.414	−1	0.000	0.000	1.999	0.000	1.414
7	1	1	−1	1.000	1.000	1.000	−1.000	−1.000
8	−1	1	−1	1.000	−1.000	1.000	1.000	−1.000
9	1	−1	−1	1.000	−1.000	1.000	−1.000	1.000
10	−1	−1	−1	1.000	1.000	1.000	1.000	1.000

Number of Runs	10	Number of Factors	3
Number of Interact	5	Number of Trials	
Fixed Row	6	Fixed Column	2
List of Factor Levels	1*1 1*2 2*2 1*3 2*3		
List of Interactions			
Starting matrix generated method	Read fixed matrix only		
Standardization	No Standardization		
Type of Design Problem	Central Composite Design - blocking		

Figure 12.23 Construction of a 10-run 3 factors central composite design with interactions.

Model robust design

An example in GO!

A project conducted recently at a major automotive company was motivated by the need to reduce the leakage of a clutch slave cylinder. Four factors were identified to be potentially significant: body inner diameter, body outer diameter, seal inner diameter, and seal outer diameter. The budget allowed for eight runs of the experiment. The engineers believed that some interactions would be active. Although it was felt very possible that two interactions are important, the engineers did not have prior knowledge on which interactions were more likely to be significant. Thus, the set of possible models included all main effects plus 2 two-factor interactions. How do we find a good eight-run factorial design that performs well over all possible models? (Li and Nachtsheim, 2000)

Number of Runs	8	Number of Factors	4
Number of Interact	2	Number of Trials	
Fixed Row		Fixed Column	
List of Factor Levels		2 2 2 2	
List of Interactions			
Starting matrix generated method		Completely randomize	
Standardization		No Standardization	
Type of Design Problem		Model Robust Design	

Figure 12.24 Construction of an 8-run and 4 factors design with interactions.

Setting of GO! (Figure 12.24):

Explanation:

This is an example of "Model Robust Design." We know in advance that there are two interactions active, but we don't know which they are. So, we can set "Number of Interact" as 2, but we can't set "List of Interactions." In such a case, a set of designs with all possible two 2-factor interactions will be generated. The criterion is modified as the sum of the whole set of designs. The CP algorithm is modified as the restricted pairwise exchange algorithm. The type of design problem should be "Model Robust Design" now.

Table 12.11 Examples with 12 runs and 5 factors g=1

Run = 12	Factor = 5		g = 1	
1	−1	1	−1	−1
−1	−1	−1	1	−1
−1	−1	1	−1	1
1	−1	−1	1	1
1	1	−1	−1	−1
1	1	1	1	1
−1	1	1	1	1
−1	1	−1	1	−1
−1	−1	−1	−1	1
1	−1	1	1	−1
1	1	−1	−1	1
−1	1	1	−1	−1

ec = 1.000000 ic = 0.9437221

Table 12.12 Examples with 12 runs and 5 factors g = 2

Run = 12	Factor = 5		g = 2	
1	−1	1	−1	1
−1	−1	−1	1	1
1	−1	1	1	−1
−1	−1	1	−1	−1
1	1	1	−1	−1
−1	1	1	1	1
−1	1	−1	−1	1
1	−1	−1	−1	−1
−1	−1	−1	−1	−1
1	1	−1	1	1
1	1	1	1	1
−1	1	−1	1	−1
ec = 1.000000 ic = 0.9203984				

Table 12.13 Examples with 12 runs and 5 factors g = 3

Run = 12	Factor = 5		g = 3	
1	1	1	−1	1
−1	−1	1	1	−1
1	−1	1	1	1
−1	−1	1	−1	1
1	1	−1	−1	1
1	1	−1	1	−1
1	−1	1	−1	−1
−1	1	1	−1	−1
−1	1	−1	1	1
−1	1	−1	1	−1
1	−1	−1	1	1
−1	−1	−1	−1	−1
ec = 1.000000 ic = 0.8625739				

More examples

Here are more examples of model robust designs. In the following examples, the design parameters are:

Run – the number of runs in the experiment
Factor – the number of factors involved
g – the number of interactions

The matrices are optimal designs found by restricted CP algorithm.

The criteria are:

ec – estimation capacity
ic – information capacity

All the calculations are done on Unix (Tables 12.11 through 12.47).

Table 12.14 Examples with 12 runs and 5 factors g = 4

Run = 12	Factor = 5		g = 4	
1	1	1	−1	1
−1	1	1	1	−1
−1	1	−1	1	−1
−1	−1	1	−1	1
−1	1	−1	−1	−1
−1	−1	1	1	−1
1	−1	−1	−1	1
−1	1	−1	−1	1
1	−1	1	−1	−1
1	−1	−1	1	−1
1	−1	1	1	1
1	1	−1	1	1
ec = 0.9952381 ic = 0.7845137				

Table 12.15 Examples with 12 runs and 5 factors g = 5

Run = 12	Factor = 5		g = 5	
−1	−1	1	1	1
−1	−1	−1	1	−1
−1	1	−1	1	1
1	−1	1	1	−1
1	1	−1	1	−1
1	1	−1	−1	1
−1	1	1	−1	1
−1	1	1	−1	−1
1	−1	−1	1	−1
1	−1	1	−1	1
−1	−1	−1	−1	1
1	1	1	−1	−1
ec = 0.9603174 ic = 0.6639671				

Table 12.16 Examples with 12 runs and 6 factors g = 1

Run = 12	Factor = 6		g = 1		
1	−1	−1	1	1	−1
1	1	1	−1	−1	−1
−1	−1	1	−1	−1	1
1	1	−1	−1	1	1
−1	1	−1	−1	1	−1
1	−1	−1	−1	−1	1
1	−1	1	1	1	1
−1	−1	1	−1	1	−1
−1	1	1	1	1	1
1	1	1	1	−1	−1
−1	−1	−1	1	−1	−1
−1	1	−1	1	−1	1
ec = 1.000000 ic = 0.9291610					

Table 12.17 Examples with 12 runs and 6 factors g = 2

Run = 12	Factor = 6		g = 2		
−1	1	1	−1	1	−1
1	−1	−1	1	1	−1
1	1	1	1	−1	−1
1	1	−1	−1	−1	−1
−1	1	−1	1	1	1
−1	−1	1	1	1	1
−1	−1	−1	−1	1	1
1	1	1	−1	1	1
1	−1	−1	1	−1	1
−1	−1	−1	1	−1	−1
1	−1	1	−1	−1	−1
−1	1	1	1	−1	1
ec = 1.000000 ic = 0.9070455					

Table 12.18 Examples with 12 runs and 6 factors g = 3

Run = 12	Factor = 6		g = 3		
−1	−1	1	−1	1	−1
1	1	1	1	1	1
1	1	1	−1	−1	−1
1	−1	−1	1	1	−1
−1	1	−1	−1	−1	−1
−1	1	−1	−1	1	−1
1	−1	1	1	−1	1
−1	1	1	1	−1	−1
1	−1	−1	−1	1	1
−1	−1	1	−1	−1	1
1	1	−1	1	−1	1
−1	−1	−1	1	1	1
ec = 1.000000 ic = 0.7978951					

Table 12.19 Examples with 12 runs and 6 factors g = 4

Run = 12	Factor = 6		g = 4		
−1	1	−1	−1	−1	−1
1	−1	1	1	1	1
−1	−1	1	1	−1	−1
1	1	1	1	−1	−1
−1	1	−1	1	1	1
−1	1	1	1	1	−1
1	1	−1	−1	1	1
1	−1	−1	−1	−1	1
1	−1	−1	1	−1	−1
−1	−1	1	−1	1	1
−1	1	1	−1	−1	1
1	−1	−1	−1	1	−1
ec = 0.9831502 ic = 0.7079295					

Table 12.20 Examples with 12 runs and 6 factors g=5

Run = 12	Factor = 6		g = 5		
1	−1	1	−1	1	−1
1	1	−1	1	−1	−1
−1	−1	−1	1	1	1
1	−1	1	−1	−1	1
1	1	1	−1	−1	−1
−1	1	1	−1	1	−1
−1	−1	−1	−1	−1	1
−1	−1	1	1	−1	1
−1	−1	−1	1	1	−1
1	1	−1	−1	1	1
1	1	1	1	1	−1
−1	1	−1	1	−1	1

ec = 0.8065267801 ic = 0.4787804186

Table 12.21 Examples with 12 runs and 7 factors g=1

Run = 12	Factor = 7		g = 1			
1	−1	1	1	−1	1	−1
−1	−1	1	−1	−1	1	1
1	−1	−1	−1	−1	−1	1
1	1	1	−1	1	1	1
−1	1	1	−1	−1	−1	−1
−1	1	−1	1	−1	1	−1
−1	−1	−1	1	1	1	1
1	−1	1	1	1	−1	−1
1	1	−1	1	−1	−1	1
−1	−1	−1	−1	1	−1	−1
−1	1	1	1	1	−1	1
1	1	−1	−1	1	1	−1

ec = 1.000000 ic = 0.9138367

Table 12.22 Examples with 12 runs and 7 factors g=2

Run = 12	Factor = 7			g = 2		
−1	1	−1	−1	−1	1	−1
−1	1	1	1	−1	−1	1
1	−1	1	−1	1	−1	1
1	−1	1	1	−1	−1	−1
−1	1	1	1	1	1	−1
−1	−1	−1	1	1	−1	−1
1	1	−1	−1	1	−1	−1
1	1	1	−1	−1	1	1
1	−1	−1	−1	1	1	−1
−1	−1	1	1	−1	1	1
−1	−1	−1	−1	1	−1	1
1	1	−1	1	−1	1	1
ec = 1.000000 ic = 0.7847008						

Table 12.23 Examples with 12 runs and 7 factors g=3

Run = 12	Factor = 7			g = 3		
−1	−1	−1	−1	1	1	1
−1	−1	1	1	1	−1	1
1	1	1	−1	−1	1	−1
−1	1	1	1	1	1	1
1	−1	1	−1	−1	1	1
1	1	−1	1	−1	1	−1
1	1	−1	1	1	−1	1
1	−1	−1	−1	−1	−1	−1
−1	−1	−1	1	1	−1	−1
−1	1	−1	1	−1	−1	1
−1	−1	1	−1	−1	1	−1
1	1	1	−1	1	−1	−1
ec = 0.9917293 ic = 0.6558642						

Table 12.24 Examples with 12 runs and 7 factors g = 4

Run = 12	Factor = 7			g = 4		
1	−1	1	−1	−1	1	−1
−1	1	1	1	−1	1	1
1	1	−1	−1	1	−1	1
1	−1	−1	1	−1	−1	1
−1	1	1	−1	1	1	1
1	−1	−1	−1	1	−1	−1
−1	1	−1	−1	−1	−1	−1
1	−1	1	1	−1	−1	−1
−1	−1	1	−1	1	−1	1
1	1	−1	1	−1	1	−1
−1	−1	−1	1	1	1	1
−1	1	1	1	1	1	−1

ec = 0.8870510 ic = 0.5201306

Table 12.25 Examples with 12 runs and 8 factors g = 1

Run = 12	Factor = 8			g = 1			
1	−1	−1	−1	−1	−1	−1	−1
1	−1	1	−1	1	1	1	−1
−1	−1	−1	1	1	1	−1	−1
1	1	−1	−1	1	−1	1	1
−1	−1	−1	−1	−1	1	−1	1
1	1	−1	1	−1	1	−1	1
−1	−1	1	1	−1	−1	1	1
1	−1	1	1	1	−1	−1	1
−1	1	−1	1	1	−1	1	−1
−1	1	1	−1	−1	−1	−1	−1
−1	1	1	−1	1	1	1	1
1	1	1	1	−1	1	1	−1

ec = 1.000000 ic = 0.8552407

Table 12.26 Examples with 12 runs and 8 factors g = 2

Run = 12	Factor = 8			g = 2			
1	−1	1	−1	1	−1	−1	1
1	1	1	1	1	1	1	1
1	1	−1	−1	−1	−1	−1	−1
−1	−1	−1	−1	−1	1	−1	−1
1	−1	−1	1	1	−1	−1	−1
−1	1	−1	1	−1	−1	−1	1
−1	1	1	1	−1	1	−1	−1
−1	−1	−1	1	1	1	1	−1
−1	1	−1	−1	−1	1	1	1
1	−1	1	1	−1	−1	1	1
1	−1	1	−1	1	1	1	−1
−1	1	1	−1	1	−1	1	1
ec = 0.9947090 ic = 0.6806318							

Table 12.27 Examples with 12 runs and 8 factors g = 3

Run = 12	Factor = 8			g = 3			
1	1	−1	−1	1	1	−1	1
1	−1	−1	1	−1	1	1	−1
1	1	1	1	1	−1	−1	−1
−1	1	1	−1	1	1	−1	−1
−1	−1	1	1	−1	−1	1	−1
1	−1	1	−1	1	−1	−1	1
−1	1	1	1	1	1	1	−1
−1	−1	−1	1	−1	1	1	1
−1	−1	−1	−1	−1	−1	−1	1
1	1	−1	1	−1	−1	−1	1
−1	−1	−1	−1	1	−1	1	−1
1	1	1	−1	−1	1	1	1
ec = 0.9227717 ic = 0.5001272							

Table 12.28 Examples with 12 runs and 9 factors g = 1

Run = 12	Factor = 9			g = 1				
−1	1	1	−1	1	1	−1	1	1
−1	1	−1	−1	−1	−1	1	1	−1
1	1	−1	−1	−1	−1	−1	−1	1
−1	−1	−1	1	1	−1	1	−1	1
−1	−1	1	1	−1	−1	−1	1	1
1	−1	−1	1	1	1	−1	1	−1
1	−1	1	−1	1	−1	1	1	1
1	1	1	1	−1	1	1	1	1
1	−1	1	1	−1	1	1	−1	−1
−1	−1	−1	−1	−1	1	−1	−1	−1
−1	1	1	1	1	−1	−1	−1	−1
1	1	−1	−1	1	1	1	−1	−1
ec = 1.000000 ic = 0.8093379								

Table 12.29 Examples with 12 runs and 9 factors g = 2

Run = 12	Factor = 9			g = 2				
1	1	1	1	1	1	−1	1	1
1	1	1	−1	−1	1	1	−1	−1
1	1	−1	−1	1	1	−1	−1	1
−1	−1	1	1	−1	1	−1	1	−1
−1	1	1	1	1	−1	−1	−1	−1
1	−1	−1	−1	1	1	1	1	−1
−1	−1	−1	−1	−1	−1	1	−1	−1
−1	−1	−1	1	−1	1	−1	1	1
−1	−1	−1	1	1	−1	1	−1	1
1	1	−1	1	−1	−1	1	1	−1
1	−1	1	−1	−1	−1	−1	−1	1
−1	1	1	−1	1	−1	1	1	⊥
ec = 0.9857143 ic = 0.5966752								

Table 12.30 Examples with 16 runs and 7 factors g = 1

Run = 16	Factor = 7			g = 1		
−1	−1	1	−1	−1	1	1
1	−1	1	−1	1	−1	−1
1	−1	1	1	−1	1	−1
−1	−1	−1	−1	1	1	−1
1	−1	−1	1	1	1	1
1	1	−1	1	−1	−1	−1
−1	1	−1	1	−1	1	1
1	1	1	−1	−1	−1	1
−1	1	1	1	1	1	−1
−1	−1	−1	−1	−1	−1	1
−1	1	1	1	1	−1	1
−1	1	−1	−1	1	−1	−1
−1	−1	1	1	−1	−1	−1
1	1	−1	−1	−1	1	−1
1	1	1	−1	1	1	1
1	−1	−1	1	1	−1	1

ec = 1.000000 ic = 0.9592438

Table 12.31 Examples with 16 runs and 7 factors g = 2

Run = 16	Factor = 7			g = 2		
−1	1	−1	−1	1	−1	−1
−1	1	−1	1	−1	−1	1
−1	−1	−1	−1	−1	−1	1
−1	−1	1	1	1	−1	−1
1	1	1	1	−1	−1	−1
1	−1	−1	−1	−1	−1	−1
1	−1	1	1	−1	1	−1
1	1	1	−1	1	−1	1
1	1	−1	−1	−1	1	1
−1	−1	−1	1	−1	1	−1
1	1	−1	1	1	1	−1
−1	1	1	1	1	1	1
−1	1	1	−1	1	1	−1
−1	−1	1	−1	−1	1	1
1	−1	−1	−1	1	1	1
1	−1	1	1	1	−1	1

ec = 1.000000 ic = 0.9111069

***Table* 12.32** Examples with 16 runs and 7 factors g = 3

Run = 16	Factor = 7			g = 3		
−1	−1	−1	−1	1	−1	1
−1	1	−1	1	−1	−1	−1
1	−1	−1	1	1	1	−1
−1	−1	−1	1	−1	−1	1
1	1	1	−1	−1	1	−1
−1	1	1	−1	1	−1	−1
1	1	−1	1	−1	1	1
1	−1	1	−1	1	−1	−1
1	−1	1	1	−1	−1	1
1	1	1	1	1	−1	−1
−1	1	1	−1	−1	1	1
1	1	−1	−1	−1	−1	1
−1	1	−1	1	1	1	1
−1	−1	−1	−1	−1	1	−1
1	−1	1	−1	1	1	1
−1	−1	1	1	1	1	−1

ec = 1.000000 ic = 0.8647436

***Table* 12.33** Examples with 16 runs and 7 factors g = 4

Run = 16	Factor = 7			g = 4		
−1	1	1	−1	1	1	−1
−1	1	−1	1	−1	1	−1
−1	−1	−1	1	1	−1	−1
1	1	1	1	1	1	1
−1	−1	1	−1	1	−1	1
1	−1	−1	1	1	−1	1
1	1	−1	−1	−1	−1	−1
−1	1	1	1	−1	−1	−1
1	1	−1	−1	1	1	1
1	−1	−1	−1	−1	−1	1
1	−1	1	−1	1	1	−1
1	1	−1	1	−1	−1	1
−1	−1	1	−1	−1	1	1
−1	−1	1	1	−1	1	−1
1	1	1	−1	−1	−1	−1
−1	−1	−1	1	1	1	1

ec = 0.9996658564 ic = 0.7620983720

Table 12.34 Examples with 16 runs and 7 factors g = 5

Run = 16	Factor = 7			g = 5		
1	1	1	1	1	−1	−1
1	−1	−1	−1	1	−1	1
−1	1	1	1	1	−1	−1
1	1	1	−1	1	−1	1
1	1	−1	−1	−1	1	1
−1	−1	−1	−1	−1	1	1
1	1	1	1	−1	1	−1
−1	−1	1	1	1	1	−1
1	−1	−1	1	−1	−1	−1
−1	−1	1	−1	−1	−1	−1
1	−1	1	−1	1	1	1
−1	1	1	1	−1	−1	1
−1	1	−1	1	−1	1	−1
−1	1	−1	−1	1	1	1
1	−1	−1	1	−1	1	1
−1	−1	−1	−1	1	−1	−1
ec = 0.9976903 ic = 0.7039213						

Table 12.35 Examples with 16 runs and 8 factors g = 1

Run = 16	Factor = 8			g = 1			
1	−1	−1	1	1	−1	−1	1
−1	1	1	1	1	−1	1	1
1	1	−1	−1	−1	1	1	1
−1	1	−1	−1	−1	−1	−1	−1
1	−1	1	−1	1	−1	1	−1
−1	−1	−1	1	−1	1	−1	1
1	1	1	−1	−1	−1	−1	1
−1	1	−1	1	1	1	1	−1
−1	−1	1	−1	−1	1	1	−1
1	−1	1	1	−1	1	1	1
1	−1	−1	−1	1	1	−1	−1
−1	−1	−1	−1	1	−1	1	1
1	1	−1	1	−1	−1	1	−1
−1	−1	1	1	1	−1	−1	−1
1	1	1	1	−1	1	−1	−1
−1	1	1	−1	1	1	−1	1
ec = 1.000000 ic = 0.9446515							

Table 12.36 Examples with 16 runs and 8 factors g = 2

Run = 16	Factor = 8			g = 2			
-1	-1	-1	-1	1	1	1	-1
-1	-1	-1	-1	-1	-1	1	1
-1	1	-1	1	1	-1	1	1
1	1	1	-1	-1	-1	1	-1
1	-1	1	1	1	1	1	1
1	1	-1	-1	-1	1	-1	1
1	-1	1	-1	-1	1	-1	-1
1	1	1	1	-1	-1	-1	1
1	-1	-1	1	1	-1	1	-1
-1	-1	1	1	-1	1	-1	1
1	1	-1	-1	1	-1	1	1
-1	1	1	1	-1	1	1	-1
-1	-1	-1	1	-1	-1	-1	-1
-1	-1	1	-1	1	-1	-1	-1
1	1	-1	1	1	1	-1	-1
-1	1	1	-1	1	1	-1	1

ec = 1.000000 ic = 0.8813823

Table 12.37 Examples with 16 runs and 8 factors g = 3

Run = 16	Factor = 8			g = 3			
1	1	1	1	-1	1	-1	-1
1	-1	-1	-1	1	1	-1	-1
-1	-1	-1	-1	-1	-1	1	-1
1	1	1	1	-1	-1	1	1
-1	1	1	1	1	1	-1	1
-1	-1	1	1	1	1	1	-1
-1	-1	-1	1	-1	-1	-1	-1
1	-1	1	1	1	-1	-1	-1
1	-1	-1	1	1	-1	1	1
-1	-1	1	-1	1	1	1	1
1	-1	1	-1	-1	1	1	1
-1	1	1	-1	1	-1	-1	-1
-1	1	-1	-1	1	-1	1	1
-1	1	-1	-1	-1	1	-1	1
1	1	-1	1	-1	1	1	-1
1	1	-1	-1	-1	-1	-1	1

ec = 1.000000000 ic = 0.7937064767

Table 12.38 Examples with 16 runs and 8 factors g = 4

Run = 16	Factor = 8			g = 4			
−1	1	−1	1	1	1	1	−1
1	1	1	1	1	−1	−1	−1
1	1	1	1	−1	1	1	1
1	−1	1	−1	−1	−1	−1	−1
−1	−1	−1	−1	1	1	1	1
1	−1	−1	1	1	−1	1	1
−1	1	−1	1	−1	1	−1	1
−1	−1	−1	−1	1	−1	−1	−1
−1	−1	1	1	−1	−1	−1	1
1	−1	1	1	−1	1	−1	−1
−1	1	−1	−1	−1	−1	1	1
1	1	−1	−1	1	1	1	−1
1	1	1	−1	1	−1	1	−1
−1	−1	1	1	−1	−1	1	−1
1	−1	−1	−1	−1	1	−1	1
−1	1	1	−1	1	1	−1	1
ec = 0.9993162155 ic = 0.7506191134							

Table 12.39 Examples with 16 runs and 8 factors g = 5

Run = 16	Factor = 8			g = 5			
−1	−1	−1	−1	1	−1	1	1
1	1	1	−1	−1	1	1	−1
1	1	1	−1	−1	−1	−1	1
−1	−1	−1	1	1	−1	−1	1
1	1	−1	−1	−1	−1	1	1
1	−1	1	−1	−1	−1	−1	1
1	−1	−1	−1	−1	1	1	−1
−1	1	−1	−1	−1	1	1	1
−1	1	1	1	1	−1	−1	−1
−1	1	1	1	−1	1	−1	−1
−1	−1	−1	−1	1	−1	1	−1
1	1	−1	1	−1	−1	−1	−1
1	−1	1	−1	1	1	−1	1
−1	−1	1	−1	−1	1	1	1
1	1	1	1	1	1	1	−1
1	−1	−1	1	−1	−1	1	1
1	−1	1	−1	1	1	1	1
−1	1	−1	−1	1	1	1	−1
ec = 0.9985569715 ic = 0.6666116118							

Table 12.40 Examples with 16 runs and 9 factors g = 1

Run = 16	Factor = 9		g = 1					
1	1	−1	1	−1	−1	1	1	1
1	1	−1	−1	1	1	−1	−1	1
−1	1	1	−1	−1	1	−1	−1	1
1	1	1	−1	−1	1	1	1	−1
−1	1	−1	1	1	−1	−1	−1	−1
1	−1	1	−1	1	−1	−1	1	1
−1	−1	1	1	−1	−1	−1	1	1
−1	−1	−1	−1	−1	1	−1	1	−1
1	−1	−1	1	1	1	1	−1	1
1	−1	1	−1	−1	−1	1	−1	−1
1	1	1	1	1	−1	−1	1	−1
−1	1	1	1	1	1	1	−1	1
−1	−1	1	1	1	1	1	1	−1
−1	1	−1	−1	1	−1	1	1	−1
−1	−1	−1	−1	−1	−1	1	−1	1
1	−1	−1	1	−1	1	−1	−1	−1
ec = 1.000000 ic = 0.9015058								

Table 12.41 Examples with 16 runs and 9 factors g = 2

Run = 16	Factor = 9		g = 2					
−1	−1	−1	1	1	−1	1	−1	−1
1	1	−1	−1	1	1	−1	1	1
1	1	1	1	1	−1	−1	−1	−1
−1	−1	1	−1	1	−1	−1	1	−1
1	−1	1	1	1	1	1	1	−1
1	−1	−1	1	−1	1	1	−1	1
−1	1	−1	−1	1	−1	1	1	1
1	−1	−1	−1	−1	−1	−1	−1	1
−1	−1	−1	−1	−1	1	1	1	−1
−1	1	1	−1	−1	1	1	−1	1
−1	−1	1	1	−1	1	−1	1	1
1	−1	1	−1	1	1	−1	−1	1
−1	1	−1	1	−1	−1	−1	−1	−1
1	1	1	1	−1	−1	1	1	1
−1	1	−1	1	1	1	−1	1	−1
1	1	1	−1	−1	−1	1	−1	−1
ec = 1.000000 ic = 0.8243225								

Table 12.42 Examples with 16 runs and 9 factors g=3

Run = 16	Factor = 9				g = 3			
1	−1	−1	−1	−1	1	−1	−1	−1
1	−1	1	1	−1	−1	1	−1	1
−1	1	−1	1	1	1	1	−1	−1
1	−1	1	1	−1	−1	−1	1	1
−1	1	1	1	−1	−1	1	1	1
−1	−1	1	−1	1	1	1	−1	−1
1	1	−1	−1	−1	1	1	1	−1
1	−1	−1	1	1	1	1	1	1
−1	−1	−1	−1	1	1	−1	1	1
1	1	1	−1	1	−1	1	−1	1
−1	1	1	1	1	1	−1	1	−1
1	1	−1	1	−1	−1	−1	−1	1
−1	−1	1	−1	−1	−1	1	1	−1
−1	1	−1	−1	1	−1	−1	−1	−1
−1	1	1	−1	−1	1	−1	−1	1
1	−1	−1	1	1	−1	−1	1	−1
ec = 1.000000000	ic = 0.7500845194							

Table 12.43 Examples with 16 runs and 9 factors g=4

Run = 16	Factor = 9				g = 4			
−1	−1	−1	−1	1	−1	1	1	1
1	1	−1	−1	1	1	−1	1	1
1	−1	−1	−1	−1	1	−1	−1	−1
1	1	−1	−1	1	1	1	−1	−1
−1	−1	1	1	1	−1	1	−1	−1
−1	−1	1	1	−1	−1	−1	−1	1
1	−1	−1	1	−1	−1	−1	1	1
1	−1	1	1	1	1	−1	−1	−1
−1	1	1	1	−1	1	−1	−1	−1
−1	−1	−1	1	−1	1	−1	1	1
−1	1	−1	−1	−1	−1	1	−1	1
−1	1	1	−1	1	−1	−1	1	−1
−1	1	1	1	1	1	1	1	1
1	1	−1	1	−1	−1	1	−1	−1
1	1	1	−1	1	−1	1	1	1
1	−1	1	−1	−1	1	1	1	−1
ec = 0.9985569715	ic = 0.6666116118							

Table 12.44 Examples with 16 runs and 9 factors g = 5

Run = 16	Factor = 9			g = 5				
1	−1	−1	−1	1	1	1	−1	−1
−1	1	−1	−1	−1	1	−1	1	1
1	−1	1	−1	1	−1	−1	−1	−1
1	−1	−1	1	1	1	−1	1	1
−1	1	1	−1	−1	1	−1	−1	−1
−1	−1	−1	1	−1	1	1	−1	1
−1	−1	−1	−1	1	−1	1	1	−1
1	1	1	1	1	−1	1	1	−1
1	1	−1	−1	1	1	−1	−1	1
1	1	−1	1	−1	−1	1	1	−1
−1	1	1	1	1	1	1	1	1
1	1	1	1	−1	−1	−1	−1	1
1	−1	1	−1	−1	−1	1	1	1
−1	−1	1	1	1	−1	1	−1	−1
−1	1	−1	−1	−1	−1	−1	−1	1
−1	−1	1	1	−1	1	−1	1	−1
ec = 0.9801348448 ic = 0.5967888236								

Table 12.45 Examples with 16 runs and 10 factors g = 1

Run = 16	Factor = 10			g = 1					
1	1	−1	1	1	−1	1	1	1	1
−1	−1	−1	1	1	−1	1	1	−1	1
1	1	−1	−1	−1	1	−1	−1	1	1
1	1	−1	1	−1	1	1	1	−1	−1
−1	1	−1	−1	−1	−1	1	−1	−1	−1
−1	1	1	−1	1	1	1	1	1	−1
−1	−1	−1	−1	−1	1	1	1	1	1
−1	−1	−1	1	1	1	−1	−1	−1	−1
−1	−1	1	1	−1	−1	−1	−1	1	−1
−1	1	1	1	1	1	−1	−1	1	1
1	1	1	−1	1	−1	−1	1	−1	−1
1	1	1	1	−1	−1	−1	−1	−1	1
1	−1	−1	−1	1	−1	−1	1	1	−1
1	−1	1	1	−1	1	1	−1	−1	−1
1	−1	1	−1	1	−1	1	−1	1	1
−1	−1	1	−1	−1	1	−1	1	−1	1
ec = 1.000000 ic = 0.8652677									

Table 12.46 Examples with 16 runs and 10 factors $g=2$

Run = 16	Factor = 10		$g=2$						
1	1	−1	−1	1	1	1	1	−1	1
1	1	−1	1	1	1	−1	−1	1	−1
−1	−1	−1	−1	1	1	−1	−1	1	1
1	−1	1	−1	1	−1	1	−1	−1	1
1	−1	−1	−1	−1	1	1	1	1	−1
−1	1	1	1	1	−1	−1	−1	−1	1
−1	−1	−1	1	−1	−1	−1	1	−1	−1
−1	1	1	1	1	1	−1	1	1	−1
1	1	−1	1	−1	−1	1	−1	1	1
−1	1	1	−1	−1	1	1	−1	−1	−1
1	−1	1	1	−1	1	−1	−1	−1	1
−1	−1	1	1	−1	−1	1	1	1	1
1	1	1	−1	−1	−1	−1	1	−1	−1
−1	1	−1	−1	−1	−1	−1	1	1	1
−1	−1	−1	1	1	1	1	−1	−1	−1
1	−1	1	−1	1	−1	1	1	1	−1
ec = 1.000000 ic = 0.7980486									

Table 12.47 Examples with 16 runs and 10 factors $g=3$

Run = 16	Factor = 10		$g=3$						
1	1	1	−1	1	−1	1	−1	1	−1
−1	1	−1	1	−1	1	1	1	1	−1
1	1	1	1	−1	−1	−1	1	−1	1
−1	−1	−1	−1	1	1	1	1	1	1
−1	−1	1	−1	−1	1	1	−1	−1	1
1	1	−1	−1	1	1	1	1	−1	1
1	−1	1	−1	1	1	−1	−1	1	1
−1	1	1	1	1	−1	1	1	−1	−1
−1	1	−1	−1	−1	1	−1	−1	1	−1
1	−1	−1	−1	−1	−1	−1	1	1	1
−1	−1	1	−1	−1	−1	−1	−1	−1	−1
−1	−1	−1	1	1	1	1	1	−1	−1
1	−1	1	1	−1	1	−1	1	1	1
1	1	−1	1	1	−1	−1	−1	−1	−1
1	−1	1	1	−1	−1	1	−1	1	−1
−1	1	−1	1	1	−1	−1	−1	−1	1
ec = 0.9999295473 ic = 0.6536532640									

The calculation time to find the optimal design is one of the essential problems in model robust designs. After applying restricted CP algorithm, the calculation time is enormously reduced. The Java applet *GO!* is capable of calculations required by model robust designs.

References

Li, W., (1997) "Optimal designs using CP algorithms," *Proceedings for 2nd World Conference of the International Association for Statistical Computing*, 130–139, 1997.

Li, W. and Nachtsheim, C.J., (2000) "Model-robust factorial designs," *Technometrics*, 42:4, 345–352.

Li, W. and Wu, C.F.J., (1997) "Columnwise-pairwise algorithms with applications to the construction of supersaturated designs," *Technometrics*, 39:2 171–179.

Li, L., Li, W., and Ayeni, B.J., (1999) "Optimal experimental designs with applications in industry," A joint research project between University of Minnesota and 3M, 3M Internal Publication.

Wu, C.F.J. and Ding, Y., (1998) "Construction of response surface designs for qualitative and quantitative factors," *Journal of Statistical Planning and Inference*, 71, 331–348.

Index